KRIEGIE

KRIEGIE

by

KENNETH W. SIMMONS

THOMAS NELSON & SONS

Edinburgh NEW YORK *Toronto*

Copyright, 1960, by Kenneth W. Simmons

All rights reserved under International and Pan-American Conventions. Published in New York by Thomas Nelson & Sons and simultaneously in Toronto, Canada, by Thomas Nelson & Sons (Canada), Limited.
Library of Congress Catalog Card No.: 60–6849

Kriegie is part of the
UNCOMMON VALOR REPRINT SERIES

Printed in the United States of America
May 2023

UNCOMMON VALOR SERIES EDITION
ISBN-13: 9781951682835

TO

FIRST LIEUTENANT TOM HEARNE FELKER

B-26 PILOT, NINTH AIR FORCE
KILLED IN ACTION OVER GERMANY
MY BEST FRIEND AND THE FINEST
PERSON I HAVE EVER KNOWN

Acknowledgments

I am especially grateful to the following gentlemen for their assistance in the completion of my prisoner of war memoirs, *Kriegie*.

Marshall L. Felker, Sr., of Avinger, Texas, chairman of the board of directors of Old Rockland Life Insurance Company, for giving me the necessary material support and providing me with the time necessary to write this book;

Donald Lewis (Jack) Ayers, of Avinger, Texas, area supervisor of Old Rockland Life Insurance Company, for editorial checking and research, but more important for his inspiration and faith in me;

William Arthur Barber, Jr., of Texarkana, Texas, attorney for Old Rockland Life Insurance Company, for editorial checking and research, but also more important for his inspiration and faith in me;

Gorham Munson, of New York, N.Y., for editorial assistance on the final draft.

K.W.S.

Contents

I.	"Our Target Is Mainz"	11
II.	"Pilot to Crew: Bail Out!"	18
III.	"Eight, Nine, Ten, Pull!"	25
IV.	The First Interrogation	30
V.	Overnight Prison Stop	39
VI.	Luft Gangsters	51
VII.	In Solitary at Dulag Luft	55
VIII.	Red Cross Packages Mean Heaven	75
IX.	The Worst Railroad Journey of My Life	82
X.	Initiation into Stalag Luft III	87
XI.	A Freshman Kriegie	103
XII.	Surprises	112
XIII.	The Goons Search the Camp	119
VIX.	Operation George	126
XV.	Rumors of Christmas	136
XVI.	The Great Bash	144
XVII.	The Escape	151
XVIII.	The Russians Draw Near	163
XIX.	The March of Death	174
XX.	The Halt at Muskau	191

XXI.	*Hell in a Cattle Car*	205
XXII.	*The Latrine Revolt*	216
XXIII.	*The Sweat-Out*	228
XXIV.	*The Battle of Mooseberg*	236
	Epilogue	247

I

"Our Target Is Mainz"

CAPTAIN MAUCK, THE OPERATIONS OFFICER, PARKED HIS JEEP IN front of barracks 21 at Heathel and lit a cigarette. He paused in the doorway of room three, and inhaled a long drag from his burning Chesterfield. Then he dropped his cigarette and ground it out. Giving a flick of the light switch, he entered the room in his usual sarcastic manner.

"All right, you bastards, first call, hit the deck, on your feet, everybody up. Let's go! Let's go!"

He crossed the room, thoroughly shaking each of its occupants. Gene Coletti sat up in his bunk, rubbed his eyes, glanced at his watch. Four-twenty A.M. Such an early awakening could mean but one thing: a long, hazardous mission deep into enemy territory.

Gene glanced across the room at me, reluctantly pushing the mass of blankets from my head. I looked in the mirror beside my bunk and saw the expression of one awakened from a trance. I rose to a sitting position, and proceeded to mount the floor.

When Terrell opened his eyes, I tried some ill-timed facetiousness.

"Roll on, thou dark and deep blue flak, roll. Just think, my friends, we have been selected once again to fly five miles above the earth, in weather only thirty degrees below zero, and we will be alone with all of those beautiful black puffs around our plane. It's so exciting, I can hardly wait."

Every night before a mission, the four officers of *Betsy II*—Gene, Tim, Terrell, and I—shaved, bathed, and arranged our clothes so that minimum time would be required in dressing. Each of us now dressed hurriedly, but was careful to re-check his clothing to see that no information of a military nature remained on his person.

The four of us passed through the mess hall door, and took our places in the line for the morning mess. The meal was cafeteria style. We moved through the main dining room to a small table in the corner. The breakfast was good. . . .

We caught the last truck for the briefing at group operations.

The truck came to a quick stop in front of group operations, and as we climbed down, Captain Mauck was standing in the main entrance checking his squadron roll.

"Your enlisted men have been inside for ten minutes, and briefing starts in two minutes. Get going," he shouted.

Terrell drew the necessary maps, while I gathered our clipboards, computers, bombing and navigation equipment. We took our seats with the six enlisted men of the crew and exchanged greetings.

Some four hundred crew members of forty heavy bombers were assembled in the main briefing room. Every man had to understand every detail of the general plan of attack, as well as his specific duties. The plan had been formulated during the night. It was now ready to be transmitted to those who must carry it out.

Briefing started with the routine order to stop smoking. Two large maps were lowered from the front wall. Blinds were drawn and the lights turned on. Major Dunn, Senior Operations Officer, mounted the stage.

"At ease. Good morning, gentlemen. This mission is of critical importance, so please give me your undivided attention."

As Major Dunn raised his yardstick and pointed to the map, whistles and groans could be heard throughout the room.

"As you all know, this red circle on the map indicates the target for our mission. As many of you have already guessed, from the early morning briefing and from developments of the past few days, our target is Mainz, Germany. It is located in this position. Mainz is located near the Rhine River—here—and only thirty miles from Frankfurt, Germany—here. The Germans are rehabilitating large numbers of scattered troops escaped from General Patton's encirclement. They are being rearmed and re-equipped into new divisions at this point just outside of Mainz.

"Our job will be to destroy the entire camp, supply area, and the city. An estimated one hundred thousand German troops are in this camp now. Equally important is the City of Mainz. It is the supply center for the entire Rhineland area of the German army. To complete this task, headquarters has ordered a maximum effort of the entire eighth airforce. The first and third divisions will attack the city while we of the second division must destroy the camp."

He paused, and then continued in a more demanding tone.

"Pilots, bombardiers, and navigators, please listen carefully from here on out. We are the first division over the target. We will be the first division to leave the coast and our group will follow the First Combat Wing. We are the lead group of the Second Combat Wing and we will bomb in group formation."

More whistles sounded through the briefing room.

"At ease," commanded Major Dunn. "The 567th Bomb Squadron will lead the attack with Jay-Bar as lead ship and Kay-Bar as deputy lead."

Terrell whispered, "I'm a son-of-a-bitch, that's us."

"At ease back there," shouted Major Dunn. "Special formation procedure will be given the pilots and bombardiers at their separate briefings. We will leave the coast at this point, Great Yarmouth being fifty miles on our left. The altitude at departure will be twelve thousand feet. We will continue climbing, crossing the French coast eight miles south of Normandy. Your altitude at this point should be seventeen thousand five hundred feet.

"You will continue climbing and reach your bombing altitude eight minutes before you reach your Initial Turning Point. The Initial Point will be this large marshalling yard on the west bank of the Rhine River. Bombing altitude will be twenty-five thousand five hundred feet. You will level off flying straight and level to the marshalling yard. You will turn 47° left as you cross this point of the marshalling yard. You will then be on course to the target. The bomb run will be five minutes long. All ships will drop bombs on the lead ship. At bombs away you will make another 47° turn to the left and fly this heading for three minutes.

"At this point you should be in the center of these three German villages. They form a triangle as you can see and they should be easily located. You will then make your last turn of 95° left which will take you directly back to Normandy Beach. The numerous check points at turns, check points en route, the time checks, air speed, altitude, and temperature will be given to the pilots, bombardiers, and navigators at their special briefings. Let me stress once again that this mission is of critical importance and a 'must destroy' target for the successful support of our ground forces. Are there any questions?"

One of the squadron commanders rose.

"Sir, will we use the same heading from Normandy back to base as we used on the map coming over?"

"Yes, Captain. Sorry I overlooked that. The same heading reversed will be used on the return leg over the channel as we used coming over. Are there any more questions? Very well, the intelligence officer will now take over."

Pilots and bombardiers were busy copying data from the blackboards while navigators outlined the mission on their maps in red pencil.

The intelligence officer climbed on the stage and stood alongside a large map hanging from the front wall.

"At ease, gentlemen. As you can see from the blackboard, takeoff is scheduled for 06:50. You will be assembled in group formation at 08:15, making your last swing over the formation check point. You will leave the English coast at 08:58. You will cross into enemy territory at 09:47 and turn on the I.P. at 11:06. Bombs away will be at 11:11. You will be out of enemy territory at 12:46 and you should be home at 13:25. Flak may be expected as scattered medium en route to the I.P. Over the I.P. on the Rhine you may expect heavy concentrated flak. From the I.P. to the target you may expect medium heavy. Over the target you may expect heavy accurate. It should be light scattered to no flak on the return leg.

"Intelligence urges you to be especially observant over the target area and record anything you might see. Recheck any personal belongings you may be carrying and wear your dog tags. Carrying A.G.O. passes is now recommended and carrying forty-five automatics is still optional. We are now advising strongly against the wearing of forty-fives. For the past three weeks the Germans have been shooting all captured allied airmen carrying arms. Let me leave you with these words. If it becomes necessary for you to abandon ship and there is any possible way to prevent it, do not bail out over the target . . . Good luck."

The meteorologist stepped on the platform and asked for the lights to be turned out. A special drawing was flashed on the screen, outlining possible bad weather to and from the target. He gave the weather predictions and concluded:

"The freezing temperature will be reached at 16,500 feet. The temperature at bombing altitude will be thirty-seven below zero. You may expect light to no overcast over the target and light to medium on the return leg. As a whole, the weather will be excellent."

As he descended from the stage, the Flight Surgeon, Major Clark, stepped to the front.

"At ease, gentlemen. Remember to check your oxygen and keep checking it all the way. Let me urge you once again to start using oxygen from ten thousand feet on up. If you leave your station to move to another station on the plane, be sure to attach a walk-around bottle to your mask. Don't get the idea you won't need it. You can't breathe without oxygen and there's no way to get around it. Nine out of every ten men who collapse from the lack of oxygen are men who thought they wouldn't need it. You all know enough about first aid to help any man who is wounded. So please help us and use it. When you return today we will have the usual double shot of bourbon for each of you."

Major Dunn returned to the platform.

"Are there any further questions? O.K. That's it, gentlemen. You are dismissed for your special briefings. Good luck and good hunting."

Conversation was noisy as the large group departed to the smaller rooms for their special briefings.

The bombardiers and navigators were formed in the East and West wings of the building. The gunners and engineers reported to the group gunnery officer, while the pilots and co-pilots remained in the main briefing room.

After some thirty minutes the small groups were dismissed to prepare for takeoff.

By 0600 all crews had gathered in the huge locker room preparing and dressing for the mission. Heavy flying clothes and boots were worn over the uniform. Earphones, leather helmets, electrically heated suits, gloves, and shoes were checked to see that each functioned properly. Parachutes were fitted and adjusted for size. Bombardiers and navigators had extra bundles to carry, composed of maps, charts, computers, clipboards, photos and other items. The co-pilots were always dismissed early to pick up the crews' lunch.

During these hurried minutes, the chaplains were on hand, helping the men dress and trying to cheer them up. Jewish, Catholic, and Protestant services were held in the locker room just before departure—short services of two or three minutes for those who desired them. The Protestant chaplain on this morning of October 19, 1944, concluded with the Twenty-third Psalm.

"The Lord is my shepherd, I shall not want. He maketh me to lie down in green pastures. . . ."

As his voice faded, some thirty trucks roared across paved runways transporting crews and heavy equipment to the parking areas, where the planes were standing.

When Terrell and I arrived at *Betsy's* side, we found the gunners on hand checking their turrets. Terrell placed the flak suits and helmets in the nose, alongside our equipment. He pinned the maps to the navigation table and carefully checked the nose instruments. I started checking the bomb load. I made sure that each bomb was locked safely to its shackle and that each fuse was screwed tight and set correctly.

Gene revved up the big engines testing each individually and he continued checking his panel instruments as Tim called off the preflight check list.

When battle stations were sounded, I checked each gunner to see that all was in readiness, and I returned to the flight deck for takeoff.

Each B-24 Liberator moved into position like a mammoth of splendid grace. *Betsy II* was third in the lineup for takeoff.

Slowly each silver bird crept down the taxistrips toward the main runway. At last all planes were in position. The time was 06:49. The first plane swung into takeoff position, facing the long concrete runway like a car poised on a superhighway. Captain Mauck was her skipper and his grim face was staring straight ahead. Steadily increasing throttles caused strong billowing winds, loud shrieking roars, and large whirlwinds of dust. Mauck watched the radio control car. The red light flashed "Ready." The huge bomber engines screamed in outrage. Seconds later the green light flashed "Takeoff." Keen eyes observed Mauck's Liberator moving instantly forward. She moved slowly at first, then faster, faster, faster, she raced down the runway, leaving its hard surface amid the dust and wind. The red light flashed again and the second bomber swung into position. "Ready." As Mauck's plane left hard earth beneath its wheels, the green light flashed again and again.

II

"Pilot to Crew: Bail Out!"

STEADILY, WITH THE UTMOST ACCURACY, THE PILOTS OF THE 389TH bomb group winged their way into formation. The group crossed its assembly point and turned as one mass unit toward the English coast. As we left the coast of England for Normandy, the 389th bomb group was out in front and the usual chatter followed.

"Leader to C-Bar, come down."

"Roger."

"Leader to J-Bar, bring your front element in closer. You are too far out. Do you hear me, J-Bar? Bring your element in and keep it there."

"Roger, sir," Coletti replied.

Betsy's code name was J-Bar. Jay was the airplane's code name and Bar was the squadron designation.

Betsy II was one of the finest ships in the group and Gene Coletti was airplane commander and master of her controls. Let us take a good look at him now, for we shall soon see him in very different circumstances. Gene was twenty-five, born and bred in New York City. He wasn't quite six feet, and was slender. He combed his black hair straight back, and had dark eyes that commanded respect. In manner he was quiet, cautious, even dignified. In one way Gene was exceptional: he was one of the few married officers I knew who remained absolutely faithful to their wives. He was the only married man in our crew of ten and we had named our plane after his wife. As commander of a lead ship, Coletti was known to be one of the best pilots in the group.

Sitting next to him in the co-pilot's seat, Tim Conway from Ohio (I think) was Gene's opposite in several respects. Tim was short, burly, heavy. He had dark curly hair, blue eyes, and an Irish face. Tim had a boisterous voice with a slight Midwestern twang. He was twenty-two.

One would think that two men as different as Gene and Tim, flying as pilots on the same ship, would never master teamwork in the air. But each was somewhat attracted by the other's character; they were both excellent pilots and, as a matter of fact, worked as a perfect team.

Terrell Hollis and I were as different as could be from the two pilots. They called us the two Southern gentlemen because Terrell was constantly talking about Mississippi, and I was forever bragging about Texas.

Terrell's number, alas, was coming up on this mission, but no one can say that he didn't enjoy his short life to the fullest. He was six feet, a couple of inches taller than I was, and outweighed me by thirty pounds; I weighed one hundred fifty and was twenty-three years old. At Mississippi State College, Terrell had been a tennis player and had never lost the deep tan acquired on the court. A handsome, dashing fellow, he was eminently successful in what he called "the endless pursuit of women."

We four officers of *Betsy* were seldom seen together socially, but as a crew we were a vibrant, well-trained, and co-ordinated team. Terrell and I, however, were real friends.

I was the bombardier and the only crew member not engaged in a fixed position. While flying to and from the target I assisted the other crew members, and from time to time I gave them a progress report on the mission. From the time of takeoff until we approached the Initial Point, I rode on the flight deck between the pilot and co-pilot. I did pilotage and dead reckoning navigation as an aid and recheck for both the pilot and navigator. Every fifteen minutes I called for an oxygen check from the entire crew, to make sure all were surviving high altitude.

This was how it went on the last mission of *Betsy II*. As we crossed the Rhine, I checked landmarks against my map. I looked at my watch and carefully made a time entry in the log. I pressed the small black button on my throat mike, and commanded:

"Bombardier to crew. Check in on oxygen."

"Nose Gunner, Roger."

That was tall Norman Alston—six foot three—blond and blue-eyed. A good nose gunner and a good party boy.

"Engineer, Roger."

Robert Bulla from Texas and always mentioning it. Robert was twenty-one, heavily built, with black hair and brown eyes. When Robert wasn't flying, he really lived it up.

"Radio Operator, Roger."

Another Texan, Charles Wheatly. Redheaded, only twenty, a bit on the quiet side but definitely not outclassed.

"Waist Gunners, Roger."

They were respectively the oldest and next to the youngest crew members. John Mather was twenty-six and the fourth member, counting me, from Texas. John was a rootin'-tootin' Texan and charmed the women with his "you-all" conversation. Poltra —I don't remember what state he came from—was twenty. He was dark in complexion, and had black hair. On an epic three-day party we had had in London, Poltra had been elected bartender and sung "I'll See You Again" as he mixed the drinks.

"Tail Gunner, Roger Dodger."

Jim Falla, making five from Texas, half the crew. He was the youngest—nineteen—and stood six feet in this crew of well-built men.

"Glad to know you all feel so damn good," I said. "Personally, I have a hangover and wish to hell this mission were over. The time is 10:16. We have crossed enemy territory and our altitude is twenty-four thousand feet. We are cruising at 164 miles per hour. The temperature is minus five degrees. We are now twenty-one minutes from I.P. and twenty-nine minutes from target. Keep calling in flak positions. When this mission is over, you can all come to my room. I'm setting up the drinks for all."

It seemed to me I was at the bombsight controls for over an hour. The nine minutes over the target were the longest nine minutes I could remember. I thought of Gene and wondered if all of us weren't getting a little flak-happy. I looked at the bombsight indices and started calling, "Nine, eight, seven, six, five, four, three, two, bombs-away! Bombs-away!"

"PILOT TO CREW: BAIL OUT!"

As the bombs hurtled from *Betsy's* side I called to Wheatly: "Bombardier to Radio, I am closing the bomb bay doors."

I reached for the bomb bay door handle.

Suddenly, out of nowhere, black puffs rose to meet us. Darkness enveloped the plane.

My hand smashed against the ceiling of the nose compartment. The altimeter spun, crushed glass separated, as shafts of air blasted inward. Pressure held all loose particles against the ceiling.

The echoing sound of rocks striking against a tin roof sang out. *Betsy II* dipped and rocked as flak slashed every part of her body. She leveled off momentarily, with her crushed nose pointed toward the horizon. I fell to the floor and Terrell Hollis came down on my head.

Then the sound of more rocks beating against a tin roof returned. Little tornadoes of air poured through huge flak holes in the nose, hurling me against the nose wheel. Everything went black.

The left wing dipped and recovered. Every part of the plane was battered by the explosions. The nose turret was crushed inward and the plexiglass nose had scattered into a thousand small pieces. The flight deck had two large holes behind the pilot's seat. The upper turret was loosened from its foundation. The bomb bay doors were blown free of the plane and the entire bomb bay section was twisted and warped. The waist was a mass of debris. The upper part of the tail was partly gone. Both wings were riddled with holes. Number two engine had been hit by a direct burst of flak and number four was on fire. Three direct bursts of flak had registered their destructive blows. In one short moment, a modern, superbly equipped bomber had been reduced to a shambles.

Terrell helped me to my feet. A quick examination disclosed that I had a piece of flak in my left leg and minor glass cuts on my forehead. Terrell was uninjured. We rejoiced to see each other alive. We removed our flak suits and fitted our parachutes tightly for the emergency ahead. Nose Gunner Alston called to the Pilot.

"Nose to Pilot. Skipper, I think I've had it. I can't move."

"Bombardier to Nose Gunner. Don't move. We are digging you out."

"Nose, Roger."

I grabbed the axe fastened above the first-aid kit on the wall and proceeded to smash the nose turret doors. When I had cut them free, Terrell dug Norman Alston from the turret. We were relieved to find no open wounds. He was suffering from shock, but nothing more.

Suddenly Gene called in on interphone.

"Pilot to crew. Prepare to abandon ship."

John Mather, right waist gunner, called in.

"Waist to Pilot. Poltra has been wounded in the left side and head. He is bleeding severely, and I have been hit in the left leg."

"Pilot to Waist. How bad are you hurt?"

"I am okay but Poltra is unconscious and looks like he is dying."

"Pilot to crew. Pilot to crew. Is anyone else injured? Check in immediately on interphone."

"Bombardier, Navigator and Nose Gunner are okay," I said.

"Radio Operator and Engineer okay," Wheatly shouted.

"Tail Gunner okay."

"Waist to Pilot. Can you land somewhere or fly back to France?"

"Pilot to Waist. I'm afraid not. We are losing altitude fast and she may blow up any minute. Number four engine is still on fire. I am sending the Engineer back to the waist. Open the camera hatch and drop Poltra through it immediately."

"Roger," Mather said.

Bob Bulla, the Engineer, was soon in the waist helping Mather. He quickly opened a first-aid kit and dressed Mather's leg. Bulla released the camera hatch door and assisted Mather in checking Poltra's chute and dragging him to the opening. Kneeling on each side of the camera hatch, Bulla and Mather lowered Poltra through the open door. Holding a firm grip on Poltra's ripcord, Bulla signalled the release and jerked hard.

"Tail to Pilot. Poltra's chute opened and he is floating down."

"Pilot to crew. Pilot to crew. Bail out! Bail out!"

When Gene uttered the command, I was standing in the nose

facing the flight deck. From an opening, parallel to my head, I could see the Pilot, Co-pilot, and Radio Operator on the flight deck above. As Gene uttered the words "Bail out," Tim Conway sprang from his seat. The cords from his heated suit and earphones were ripped from their sockets. He threw his oxygen mask to the floor and leaped toward the bomb bay section. His large buttocks could be seen on the catwalk. Facing the flight deck, he dove through the bomb bay opening into the halls of air.

Tail Gunner Falla, who was standing by the camera hatch, called to the Pilot over interphone.

"Tail to Pilot. The left Waist Gunner just bailed out and I am leaving immediately." Without waiting for recognition, Falla pulled his head set and oxygen mask free and made a swan dive through the waist camera hatch opening.

I jumped to the side window and saw the three white chutes floating downward. Time was growing short. I grabbed the red handle locking the nose wheel doors and jerked it toward me with a terrific pull. The pins released and the doors fell open. Alston was crying as I slapped him hard across the face. I grabbed him by the jacket and shouted in his ear.

"Bail out, she's going to blow up!"

Norman Alston had been without earphones since leaving the nose turret and hadn't heard Gene's orders to bail out. He crawled across the nose, taking a sitting position with his feet hanging through the opening. He turned to Terrell and me, simulated the Air Corps' handshake, and leaped feet first into the open air.

Terrell jerked his oxygen mask free and embraced me firmly. Looking me squarely in the eyes, he shouted: "Auf Wiedersehen."

He turned to the opening with his right hand on his ripcord and jumped from the standing position with his feet close together.

In the freezing air, sweat was pouring from my body. I tried to keep my senses. Pressing my throat mike button inward, I called:

"Bombardier to Pilot. Navigator and Nose Gunner have bailed out. Did the Engineer and Radio Operator get out okay? I am ready to jump. Acknowledge."

"Pilot to Bombardier. Hold on a minute. She is flying straight and level, and I got the fire out in one engine. I am going to try to make it back. Give me a straight course to France as quickly as possible."

I grabbed the maps from the floor and started figuring a new heading. I saw Gene turn around to check on Bulla and Wheatly. As I looked past Gene, my eyes froze in disbelief.

Bulla had returned from the waist section to find Wheatly draped in two flak suits. He was sitting on the edge of the flight deck looking through the open bomb bay sector. The open spaces beyond his view had paralyzed Wheatly's body with fright. He kept staring downward, holding tight to the steel brace on the wall of the plane. I saw Bulla shake him and demand that he jump, but Wheatly didn't see or hear him. Bulla raised himself onto the flight deck and grabbed Wheatly's arms. He jerked with all his strength and Wheatly's frozen hands came free. Bulla pulled both flak suits free of Wheatly's body. Kneeling behind him, the Engineer lunged forward. Wheatly fell on the catwalk. Bulla followed up with a forward thrust of the body and Wheatly cleared the bomb bay opening with hands and legs spread. Bulla then signalled Gene by a wave of his hand and jumped.

I had located our position on the ground and had mapped a new course to France. I had figured that flying a heading of 182 degrees for ninety miles would put us in allied territory. But before I had time to lay down the maps, Gene called me.

"Pilot to Bombardier. There's no hope of saving her. Bail out immediately. I am leaving."

"Bombardier to Pilot. I have the new course."

"Pilot to Bombardier. Bail out at once!"

"Bombardier to Pilot. Roger."

I picked up my forty-five automatic lying next to the intervoltmeter and fired five shots through the Norden Bombsight. The rate end and other intricate mechanisms were shattered to pieces. I threw the gun on the floor. I pulled my gear free and faced the nose. The limitless blue horizon stood before my eyes. It was unbelievable that I could be bailing out over Nazi Germany.

III

"Eight, Nine, Ten, Pull!"

I TOOK A FIRM HOLD ON MY RIP CORD, AS I MOVED FORWARD.

Harsh, shrieking winds burned my eyes and battered at my face as I crouched to jump. Cold sweat broke out in the palms of my hands and fear engulfed my mind. I wondered if I would ever see my home again.

Like a swan dive from a ten-foot diving board, I executed this dive with my best technique. Eyes closed, I left *Betsy II*, praying as I jumped.

I had bailed out of a heavy bomber flying down wind on the target at a real air speed of over 300 miles per hour. The altitude was probably somewhere around 22,000 feet—more than four miles above the earth. What I had been trained to do was to count one for each thousand feet of fall until I was 5,000 feet above the ground. That is, I was supposed to count from one to seventeen, and then pull the rip cord of my parachute to open it at approximately one mile above the earth.

The reason for the count is this. As one left the plane, it was going forward at a speed of 300 miles per hour down wind over the target. Therefore, when one jumped, one was moving forward at the rate of the speed of the plane, but gradually one slowed down to the rate of gravity, the normal rate that any given weight will fall through space. Thus, for some seconds after leaving a plane, one is going at a terrific rate of speed but gradually slowing down. That is why one should count up to sixteen or seventeen before pulling the rip cord. I didn't. I knew better but I became afraid the parachute might not open.

". . . seven, eight, nine, ten, pull," I counted, and in a panic jerked the rip cord fully forward. I was still moving at too swift a rate. A terrific concussion occurred.

My neck and shoulders felt paralyzed . . . pain streamed into my head. Suction hurled my shoes into space. My socks stayed on; I don't know why. The suspenders of my heavy flying pants popped from their buttonholes. My leather helmet came free from my head. My spine felt as if it had been crushed and my legs were benumbed.

Some strange force held a constant upward grip on my chest and armpits. I opened my eyes and looked up.

Towering above my head, I saw a huge umbrella of white silk drifting parallel with my body. Suspended from a thousand cords, I dangled in space like a victim of the hangman's noose. I sighed with relief, as I watched the white parasol floating above me.

As my head cleared, my ears detected the ever increasing roar of bombers' engines approaching in the distance. Above and to my left they came slowly forward. The city of Mainz and towering clouds of smoke were directly beneath my feet. I knew what it was. It was the Third Division's attack on the target.

As the planes approached overhead, flak rose from the ground, springing straight up to encounter and destroy the enemy. Flying through the air, hundreds of small metal chips disintegrated from the large exploding shells. Above, below, and horizontal to my body, black puffs began to form. I was directly over the target and exploding shells began to surround me. As bombs were released from the planes overhead, the heavy bombardment of flak increased.

My parachute swooped and rocked violently from the concussion of several close misses. Bells rang in my ears as a temporary state of deafness engulfed me. My eyes burned and my vision blurred. My head was again bursting with increasing throbs of pain. Clouds of darkness seemed to encircle me. Breathing became difficult and fear mounted in my body.

As death stood before me, I experienced a strange feeling and felt emotions that would forever remain a vivid picture in my memory. Knowing I was going to die, I became meek. With heavy eyes, uplifted head, and a humble heart, I called out the name of God. . . .

I again opened my eyes. Several thousand feet below and half a mile to my left, bombs were exploding in the city of Mainz. High above the city black puffs encircled the armada of planes. I had drifted clear off the target and continued to drift further away as my parachute descended toward the ground.

Looking about me, I spotted three parachutes. One was about two thousand feet below and the other two were skimming the treetops near the ground. I guessed my altitude to be four thousand feet and I realized I would reach the ground in a matter of minutes. I knew that Gene was the last man to leave the ship, but I could not locate any parachute above me. Had Gene delayed his jump? Had his parachute failed to open? Had it been destroyed by flak? Of course, there was the possibility that Gene had not been able to jump. However, I doubted this. I wondered what had happened to the other six members of my crew. Surely the Germans would not dare to kill them.

As the ground grew nearer and nearer, I tried to recall every lecture I had attended on parachute jumping. "For controlled drift of parachute, pull cords downward in the direction drift is desired." No, ". . . pull cords in the opposite direction to that in which drift is desired."

I decided I would try. I took a firm hold on as many strands as I could grasp and pulled straight down. I tried to observe the ground to determine the direction in which I was drifting. If my parachute changed its directional movement, I was unable to tell it. I tried a couple more times, but could not detect a change of direction. I was falling fast now, or at least it seemed so because I was much closer to the earth. I estimated the ground to be less than a thousand feet away and tried to improvise a hurried plan of escape.

I looked directly below and spotted a large superhighway leading into a small village. To my left, I observed another village of equal size with a small German airport located halfway between the village and superhighway. To my right were a few scattered farmhouses with vast acreage of cultivated fields. I was thinking of the instructions given us on methods of escape. The most important point to remember was to hide in a forest or wooded area during the day and travel slowly by night. I realized for the first time that we had never been told what to do if we couldn't find a forest.

I knew at once that I must miss the highway or take a chance of breaking a leg. I turned sharply to the right, grabbing the extended cords above my head, and I started to jerk. Just as suddenly, I released the cords. I remembered some important words from another lecture on parachute jumping. "Never attempt to correct drift the last two hundred feet. If your parachute should collapse, you will be too low for it to reopen."

Looking down I saw the ground less than fifty feet away. Parts of past lectures came back to me. "Keep a level head. Pick your spot to hide. Remember everything in sight. Hide your parachute. Relax." Whop!

I had cleared the highway some hundred yards to the right and hit the ground in a newly plowed field. Big gusts of wind carried the white umbrella forward, dragging me across the furrows. I tried to regain my balance, but each attempt was in vain. Clods of dirt hit my mouth and eyes. Sharp rocks cut my face and hands. My body was bruised and battered from sharp rocks and rough ground as the parachute dragged me forward. My muscles were weak and jumpy as I struggled to regain my balance. Finally a great gust of wind thrust my chute forward, entangling it among the branches of two small apple trees. I stumbled forward, freeing myself as the white umbrella collapsed. I quickly removed my Mae West and looked about.

Dead ahead, about a quarter of a mile, the small village stood. Halfway between me and the village, some sixty men, women, and children were moving toward me. Now they started to run, and I could see they were armed with spades, shovels, pitchforks, clubs, and other implements.

To my left two trucks had stopped on the large superhighway, and soldiers were rushing toward me.

Directly behind me and to my right, a dozen old farmers, fully armed, came slowly toward me in spread formation. I was completely encircled and hopelessly outnumbered.

The soldiers were less than thirty yards away when I turned to the right. My eyes caught the facial expression of one bloodthirsty farmer. Bitter eyes, cruel curved lips, and a desperate face were convincing evidence of his intentions. Realizing where my one chance for survival lay, I rushed to the soldiers and surrendered.

The soldiers formed a circle about me as the revengeful civilians drew near. The Lieutenant in charge barked orders to the civilians. Although I did not understand a word of German, I was convinced that the Lieutenant would make every attempt to take me prisoner and protect me from the growing crowd. I was immediately ushered back to the highway and shoved through the open door of a black staff car. The Lieutenant and three privates got in with me. At a command from the young German officer, the two trucks and car started toward the village. I was glad to leave the angry crowd behind.

As we neared the outskirts of the village, the car came to a stop. The right back door swung open, the Lieutenant looked at me and shouted:

"'Raus! 'Raus!"

IV

The First Interrogation

I GATHERED MY PARACHUTE AND MAE WEST AND STUMBLED ONTO the highway in my stocking feet. I tried to appear unafraid. Every muscle in my body ached, but I put on a calm face showing no sign of distress.

At a command from the Lieutenant, a guard of six soldiers formed about me. We moved off in marching formation with the young officer out in front. As we started through the village scattered civilians gathered to observe me. I saw bitter faces and wild staring eyes.

As we marched forward the crowd grew larger. Several old women raised their hands and shouted: "*Luft Gangster, Schwein, Schwein.*" Others cursed and spat as we moved on. The situation was tense, but I knew the civilians would not dare to attack German armed guards. I wondered where they were taking me and tried to review lectures on what to do if you were captured.

"Tell them nothing except your name, rank, and serial number. Be military, polite, and courteous, but at the same time be forward and firm with every statement you make. Try to remember everything you see, especially anything of a military nature."

At the second street corner, we turned right and continued in that direction for five blocks. Near the other end of the village we crossed the street on our left and passed between two small buildings into a clearing in the distance. A large carpet of deep green grass blanketed the ground before us. As we approached a long sidewalk, I felt a cooling sensation from the tall, thick pines shading our path. Some fifty yards away there was a large stone building with brown sides and green roof. I observed a driveway winding through the pines that led to the side of the building. As we approached the building the pines grew much thicker in number.

As we reached the front steps an elderly officer dressed in light brown uniform and shiny brown boots, knee high, stepped forward to meet us. He wore a red swastika on his left arm. I had never seen a uniform of this color and could not ascribe it to any branch of the German armed forces. I knew that my guards were members of the German Air Force or *Luftwaffe*. I was also familiar with the uniforms of the *Gestapo, Wehrmacht*, and Storm Troopers or S.S. I decided the officer in brown uniform must be a member of the political government or a member of the Nazi Party.

As the elderly officer approached me, the guard standing next to me stepped aside and snapped to attention. Then the officer in brown thrust his arm forward, pushing me sideways. He shouted in my face and screamed in a hysterical manner. He acted, I felt, as if he were trying to impress the soldiers that he was a member of the super race. I stood silently at attention, and my silence increased his anger. Probably he had hoped I would try to defend myself and thus give him a reason to kill me. He continued to shout, "*Heil Hitler. Heil Hitler. Luft Gangster. Luft Gangster.*" He finally worked himself into a fanatical rage. He spat in my face and slapped me violently across the forehead. Humiliation and rage raced through my mind, but I could do nothing except remain silent.

The officer in brown grew red in the face and drew back his fist, but the young Luftwaffe officer grabbed his arm and shouted. "Nix. Nix. Prisoner."

Without waiting for an answer, the young officer commanded the guard to follow him and we proceeded to the main entrance of the building. I realized that for the first time I had come face to face with a typical Nazi fanatic.

As we reached the top of the steps, a guard on duty at the front door snapped his heels, shouted "Heil Hitler," and pulled the door open. The Lieutenant, one private, and myself went through the door while the remaining guards stayed behind. As we went down the hallway, I noticed offices on each side. Names and military rank were printed in large black letters on each door. At the end of the hallway, the Lieutenant opened a large oak door that led into a small reception room.

The room was empty except for a medium-sized desk, typewriter, and chair. There was one small window above the desk, and a door adjoining the left wall. Presently the door came open and a young blond soldier, my age, stood before us. The Lieutenant and private popped to attention and heiled Hitler. I knew he was the Lieutenant's superior officer and from the conversation I decided he was a captain. He smiled and appeared to be very friendly as he motioned for me to follow.

His office was nicely furnished. He motioned for me to sit down, and we studied each other, as he babbled to me in German. Realizing that I did not understand, he reached for a small box on his desk and offered me a Turkish cigarette. I conveyed an expression of gratitude as the youthful captain produced a match. I inhaled three long drags before withdrawing it from my mouth. I had never enjoyed a cigarette more in my life.

The Captain picked up a French-style telephone and began a long conversation in German. I sat silently, smoking and listening to the strange words spoken by the Captain. I was amazed by the elaborate furnishings of his office.

As I reached for an ashtray, the Captain finished his telephone conversation and reached for one of the books on his desk. I knew that before I got out of this building, they were going to ask many questions and try to get information from me. I again tried to review all of the lectures given by Intelligence on the conduct of a prisoner of war. I recalled two shows I had seen on how to act if you were captured by the enemy. You must give them nothing more than name, rank, and serial number. Giving any information beyond this would be considered an act of treason.

We had been taught that all German interrogators were superior to us in mental wit and that most of them were trained psychologists. We knew them to be extremely clever and capable of translating any roundabout answer to a question. They could detect a lie immediately. I knew they would use any possible means to misguide me, or to create any impression that would cause me to talk. It was pointed out time and again at these lectures that the Germans had thousands of tiny pieces of information at their disposal; one's carelessness could add a link to the chain of German intelligence.

It seemed fantastic that I could now be in a helpless situation. In the past I had always had friends in any critical hour of need. Now, I was friendless, alone. I had always placed such high value on friends, money, and position, but now I was aware that no material possession on earth could help me in this crisis. For the second time this day I knew that God was my only hope and without His help I felt I was surely doomed for destruction. Again I prayed silently and seriously.

Suddenly the office door swung open and a young German woman came forward to the Captain's desk. She was dressed in plain civilian clothes and I judged her to be less than twenty-five years of age. She was very attractive and talked to the Captain several minutes. Finally she faced me and smiled.

"Good afternoon," she said. "Do you have any papers to identify yourself?"

I removed my dog-tags from around my neck and pulled my A.G.O. pass out of my jacket. As I handed them to her, she snapped a command to the guard standing by the door. He approached me and commanded: "*'Raus.*"

As I looked at him, he pulled me from my seat.

"Please relax while he searches you," she instructed me. "You will have to undress. You may leave your shorts on."

I removed my heavy flying clothes while the private searched them thoroughly. Underneath I was wearing a uniform of pink pants and forest green shirt.

"So you are a lieutenant," she said. "I haven't seen an American officer in several years."

I could tell that she had been around. Without answering, I removed the rest of my clothes and handed them to the guard. In less than five minutes the search was over. I quickly dressed while the girl and the Captain checked my belongings. They had found one escape kit containing several thousand francs, a compass, maps, food, and first-aid equipment. Besides the kit, I was carrying my dog-tags, my A.G.O. pass, a cigarette lighter, a pack of Chesterfield cigarettes, and a box of matches. In a few minutes the girl returned my lighter and cigarettes and smiled.

"You may smoke if you desire," she said. She studied my picture on the A.G.O. pass while I fired up a cigarette.

"Your name is Kenneth W. Simmons. You are a First Lieutenant in the U. S. Army Air Force and your army serial number is 0-717115."

"Yes, ma'am, that is correct."

"How long have you been in England?"

"Sorry, I can't answer that."

"But why, I don't understand."

"I can only give you my name, rank, and serial number."

She turned to the Captain and spoke a few words in German as if she were explaining my answer. They carried on a short conversation, and again she faced me.

"I must ask you some questions, and I am afraid you must answer them," she explained. "First of all, what bomb group did you belong to?"

"I can't answer that question."

"You were a member of the 8th Air Force because you were shot down in a B-24 Liberator and you came from England. Isn't that true?"

"My name is Kenneth W. Simmons. I am a 1st Lieutenant in the U. S. Air Force, and my serial number is 0-717115."

"Your navigator is dead," she announced.

I tried to keep a noncommital look, but her statement caused considerable shock. Terrell dead? How could he be? Who would have killed him? How would she know that he was my navigator? She was probably trying to confuse me and trick me into answering some question.

"Really? How do you know he is dead?" I asked.

"We found his body when the civilians had finished with him. He was killed six kilometers this side of Mainz. He was wearing a set of navigator's wings above his shirt pocket, and he bailed out of the same plane you jumped from."

"What was his name?"

She smiled and answered: "Lieutenant Terrell Hollis. They beat him badly before he died and he told the civilians plenty. Now do you deny that he was the navigator of your plane?"

"You are wasting your time. I never heard of him before in my life."

"You are lying, Lieutenant Simmons. I saw the look on your face when I told you he was dead. He was killed by some local civilians for the brutal bombing of innocent women and children. What type of bombs were you carrying?"

"How should I know?"

"Because we are almost certain that you were a lead bombardier. Your plane was leading the formation, so we know you are an officer of a lead bomber. We have reason to believe that you are a bombardier and that you did the bombing. We want to know why you were bombing this target."

"My name is still Kenneth W. Simmons. I am still a 1st Lieutenant in the Air Force, and my serial number is still 0-717115."

"You are becoming difficult. We must identify you as a prisoner of war. If we are not convinced that you are a prisoner of war, you will be shot as a spy. What are the names of your crew members so we can identify them as military prisoners?"

"I can't answer that."

"Do you think they have any possible hope of escaping from us?"

"I don't know."

"You lived in England, Lieutenant, and you saw our great secret weapon. What did you think of the big buzz bombs?"

"I haven't given it much thought."

"So you did live in England," she shouted.

"I didn't say I lived anywhere."

"You admitted you lived there, and you know our bomb is a great success. Is it not true that half of London has been destroyed?"

"You are wasting your time."

"Answer me. Didn't the bombs destroy half of London?"

"I know nothing about bombs. I can only give you my name, rank, and serial number."

She grew very serious and looked at me with a cruel expression.

"Do you realize that the civilians would have killed you if we hadn't stopped them?"

"Yes, I suppose they would have tried to kill me."

"Don't you understand that we could still turn you over to them? If we released you in the streets of this village you would be torn to pieces. You must be identified as an American prisoner of war. You could easily be a spy. That is why we ask you these questions. You must convince us that you are a member of the American armed forces. It is useless for you to continue like this. We already know most of the answers and we are many against one. Unless you answer my questions now, we will be forced to use more drastic measures. We will go as far as you make us go. You probably don't know the meaning of hunger and torture in prison. We will give you the works if necessary, but in the end we always break them down."

She paused to light a cigarette.

"You bomb and kill our women and children. If you are expecting any kind of mercy from us, you won't get it. Why should we care about air gangsters? You are the most hated group in Germany, and unless you answer now it will be most regrettable for you. Do I make myself clear?"

I answered slowly:

"You just said you saw me bail out of a bomber. You have my dog-tags and my identification. You know I am a prisoner of war. I am also a soldier and you know I cannot betray my country. I believe as strongly in the cause for which I fight as you believe in your cause. I am only one against many, but I am also one among many Americans, and I will never do anything to betray my country. I expect to be treated as a prisoner of war, but if I am not treated as such, then I will have to make the best of it."

The young girl sprang forward and slapped my face.

"You are a Luft gangster and you kill little babies and innocent people. You will answer my questions or you will be killed by the civilians. What part of England did you come from?" she screamed.

I did not answer.

She slapped me again and screamed again.

"What part of England did you come from?"

"My name is Kenneth W. Simmons. I am a 1st Lieutenant in the Air Force and my serial number is 0–717115."

She again turned to the Captain and spoke several minutes in German. She turned back and looked at my Mae West.

"The Captain wants to know what that thing is for," she said.

"It is called a Mae West, and it is used as a life preserver if you bail out over water."

She translated this to the Captain and turned to me again.

"The Captain wants you to demonstrate it to him."

I smiled as I picked up the Mae West and reached for the air valve. As I pushed the valve in, the Mae West shrieked and jumped as it expanded with air. The Captain and the guard both drew and pointed their pistols at me. I quickly explained to the girl.

"That is the way you fill it with air before you hit the water."

As she explained it to the Captain, he came from behind his desk and examined it carefully. Suddenly he began to laugh and all of us laughed with him.

He barked a command to the guard at the door. In a few minutes the guard returned and handed me a hot cup of coffee. It was stronger than any coffee I had tasted before and I drank it in silence.

While I sipped the coffee, the girl and the Captain talked for several minutes. Finally, she turned to me and spoke in a pleasant voice.

"Do you intend to resist us longer? If you do we will send you to a place where you will learn the meaning of resistance."

"No, I don't intend to resist anyone. I am your prisoner, but I have told you all I will ever tell."

She spoke to the Captain once more. He rose to his feet and executed a Heil Hitler salute. She reached the door and turned to me.

"When we win the war you will be returned to this village to help rebuild it."

Without another word she departed.

I had been surprised by her excellent pronunciation of the English language. I kept wondering about what she had told me about Terrell.

In a few minutes the Captain offered me another cigarette. I offered him one of my Chesterfields and we both lighted up. He watched me for several moments and finally motioned for the guard to take me out. I gathered my Mae West and parachute in my arms and followed the guard to the door. The Lieutenant was waiting in the outer office and saluted the Captain as he entered. The Captain issued some commands and smiled as he departed.

At the side entrance I was ushered into a private car, greatly surprised to see that it was a Ford V-8. The Lieutenant rode in the front seat with the driver, while I slipped into the back seat with an expressionless guard.

I was glad the interrogation was over and reviewed every word of it in my mind. I was positive I had told them nothing. I wondered where they were taking me and if they would carry out the young woman's threats of punishment. She had turned out to be much smarter than I had expected her to be.

I wondered now if they would take me to Dulag Luft. Every American flyer knew of Dulag Luft as the German torture chamber for Allied fliers.

I was totally exhausted from the most tiring day of my life. My muscles ached with pain. I lay my head back on the seat and closed my eyes. My head became heavy and I fell asleep.

V

Overnight Prison Stop

THE COUNTRYSIDE WAS GREEN. TALL, SLENDER PINES GREW IN ROWS along the highway. We passed scattered German farmhouses and each appeared to be clean and well-kept. There were no signs of poverty or destruction.

Our small German staff car came to a stop at a three-way intersection and waited for the proper signal to cross. As we moved slowly forward, the Lieutenant spoke to the driver, and pointed to a large farmhouse a few hundred yards ahead.

I had been awake several minutes, but I remained motionless with my eyes half closed. They didn't seem to notice me and I was glad.

I watched the large, white, two-story, colonial-style farmhouse draw nearer. I knew we were going to stop and I wondered why the Lieutenant had selected this particular house. As our car pulled up in front, the Lieutenant stepped out and moved briskly toward the front gate. This house reminded me of something I had seen in the past. The four gables forming a part of the roof and the three white beams supporting the curved balcony of the second floor were vivid recollections. *I felt as if I had been here before.* I even remembered the white picket fence enclosing the half-acre of ground and the large tree shading the front lawn.

A small German boy, wearing the uniform of the Hitler Youth, sat in the front yard playing with a toy gun. As the Lieutenant passed through the gate, the boy sprang to his feet, thrust his right arm forward and shouted: "Heil Hitler."

The officer clicked his heels, returned the salute, and then continued to the front porch.

A middle-aged woman sat in a green rocking-chair darning stockings. She seemed unconcerned over the Lieutenant's visit and continued her darning as they talked. Then she spoke rapidly and pointed to the highway intersection. The Lieutenant executed a polite salute, and she nodded in response. As he departed, she glared at the car, trying to get a glimpse of me. Once again the small lad by the tree executed the "Heil Hitler" salute. As the Lieutenant re-entered the car, I noticed a far-off expression in the boy's eyes. I judged him to be about seven years old.

We returned to the intersection, and the driver made a sharp right turn. To the right I saw a white sign with black letters which said: "Mainz—16 Kilometers." This gave me a strange feeling. We had bombed Mainz earlier this day and I wondered if they were now taking me to the target. They might be taking me to the spot where Terrell had been killed! I wondered if Terrell were really dead and if they were going to exhibit him in some horrible condition in an effort to make me talk. I prayed that I would not be forced to witness such a sight. With each passing minute, I felt more certain that Terrell's mangled body would be used to break me down.

The girl interrogator had known that Terrell was the navigator of a B-24 Liberator. She was almost certain that I had bailed out of the same plane. If I were taken to Terrell's body, they would watch my every move and expression. I remembered that I had sworn to give up my life before I would reveal any operations of the Norden Bombsight. I knew now that if they took me to Terrell, I must face him as a stranger regardless of our close friendship. I prayed that I could.

As we raced down the highway, I looked for my wrist watch. It was gone. I had been wearing it when the plane was hit, but I never saw it after that. The band must have snapped when I bailed out. I guessed it was past four o'clock and I wondered when I would get something to eat.

The car climbed a long hill. We slowed down as the highway grew steeper; near the top, it was necessary for the driver to shift into low gear. As the countryside became visible below, I saw towering clouds of smoke in the distance. We were six to eight kilometers away from the city of Mainz, which looked like the smoldering ruins of a once great city. Huge fires still flared in the distance and extended for miles. Parts of the environs of the city seemed leveled to the ground while burning buildings were still standing in the heart of the city. I had never seen such thorough destruction.

The Lieutenant glared at me and pointed toward the clouds of smoke, shouting: "*Mainz. Schwein. Luft Gangster.*"

We rode in silence for the next ten minutes. Two miles from the city limits we turned off the highway and came to a stop at the entrance of a small air base. I observed a runway running north and south and at the end of it a "T" running east and west. Several Focke-Wolf 190's were dispersed about the parking areas on both sides of the runway. The entire camp covered some twenty acres of ground. I guessed there were around twenty buildings, counting hangars, administration buildings, barracks, and mess hall. I knew this field was a branch of the German Fighter Command, and I hoped they would keep me overnight.

The guard on duty at the front gate stepped forward to identify the car and its occupants. We passed through the main gate, turned right at the first crossroads, and parked in an open drive adjoining a single-story, gray stone building with cross bars covering the windows. This had to be the guardhouse.

A burly, red-faced, middle-aged sergeant sat in the front office, his feet propped on his desk. Seeing the young Lieutenant, he lowered his chair to the floor with a look of disgust, and executed an indolent salute from the sitting position. The Lieutenant grew red in the face, sucked in his stomach and rushed forward, ordering the Sergeant to his feet. He placed the Sergeant in a rigid brace and made him execute the "Heil Hitler" salute for several minutes. When he finished, the Lieutenant severely reprimanded the Sergeant for showing disrespect to a German officer and copied his name in a little notebook.

I enjoyed this show very much, but dared not let my pleasure be seen.

The Sergeant proceeded with a new air of efficiency. He listed my belongings, searched me, and assigned me to a cell in less than five minutes. He screamed at the private on duty in the hall and cursed him for not opening the steel door fast enough. There were five cells on each side of the hall. Apparently I was the only prisoner, for all cell doors stood open. I was escorted to cell number two, at the far right corner of the hall. The door slammed closed as I entered. Hearing the key turn from the outside, I knew that I was there until the Germans decided to remove me.

My cell was only ten feet long, six feet wide, and twelve feet high. The floor and walls were made of a gray stone and the door was solid steel with no peep holes to see in or out. For light or ventilation there was a small window, two feet square, located in the end of the room. No furnishings except a small pile of hay in the middle of the floor. The cell was filthy. Spider webs hung from the ceiling and corners. Body droppings lay on the floor below the window. Some previous prisoner had used the cell for a toilet and the scent was still there.

In all this discomfort, I felt relatively happy. I was temporarily safe from civilians as well as fanatics like the officer in brown at the interrogation center. For the first time since my capture I could relax. My legs and back felt stiff and extremely sore, as if something had crushed them. I could hardly move my neck, and even though my headache had gone, I still felt the stinging pain from several open cuts on my face and arms. A piece of flak had grazed my leg, but the wound was healing over. I felt slightly dizzy and I knew I was nearing a state of exhaustion.

I reclined on the floor and tried to put my weary body into a comfortable position. My feet were cold from standing on the damp floor. I buried them in the hay and used my heavy flying jacket for a pillow. After lying down, I fumbled in my pocket for a cigarette and match. I found I had eight cigarettes left. After careful consideration, I decided to enjoy each one of them during moments like this. I struck a match and watched the fire spring from the match head. I inhaled several long drags, I pinched the hot end of the cigarette, extinguishing its light, and placed the butt back in the pack. In a few minutes I was fast asleep. . . .

I was awakened by the rattling of a key in the door. It was dark, but as the door swung open, the ceiling light came on. A German boy about eighteen years of age stood in the doorway, neatly dressed in a light blue Luftwaffe uniform. His shoes were highly polished and his brass was spotless. He was without rank, and armed with a rifle. I was certain he was the new guard on the night shift. He had a pleasant smile and his eyes showed kindness. He stepped inside and helped me to my feet.

"*Essen.*" The expression on my face told him that I did not understand.

"*Essen mit Kamerad,*" he said, pointing to his mouth.

"Roger, understand, understand," I answered.

"*Jawohl,*" the guard laughed.

"*Jawohl,*" I repeated.

As we entered the hall, I noticed the door to cell number six was closed, and I wondered if one of my crew had been brought in. The young guard rattled the keys as he unlocked the door. I tried to restrain myself, but as the light came on, Norman Alston, the nose gunner, rushed forward and embraced me firmly. I didn't care whether or not the guard knew we were crew members. I returned the embrace and Alston cried. The guard left us standing in the doorway, but returned in a few minutes, carrying our suppers on a tray. Two bowls of soup, four slices of black bread, two cups of coffee. He placed the tray on the floor, smiled, and departed, locking the door behind him.

Alston had landed on a flat-top building in the town of Wiesbaden, about ten miles from where I had been captured. While he was getting off the roof of the building several shots were fired at him. He was beaten by a crowd of civilians in the streets, but was finally rescued and captured by the Wehrmacht. He had suffered a minor cut above the left eye; otherwise he looked as if he never felt better.

I gulped a mouthful of coffee and nearly choked.

"My achin' back! I have never tasted such slop in all my life," I said.

"They call it root water," Alston explained. "I think it was made from boiled roots. If you think the coffee is bad, wait until you bite into this compressed sawdust they call bread."

"Well, at least the soup is good. I am so damn hungry I could eat anything," I said.

"By the way, I saw Tim Conway in Wiesbaden. He was captured by some more Wehrmacht boys and when I saw him they were taking him to the town jail," Norman said.

"Well, I'm glad he is alive, but I wish to hell we knew about the others. Poltra got a piece of flak in his head. I doubt if he was alive when we dumped him out. Mather also got a bad hit in the leg."

"Damn, Kenneth, I didn't know that," Norman exclaimed.

Presently the guard returned. He carried our trays into the hall and signaled for me to follow. I returned to the lonely interior of cell number two.

Once again I retired to my bed of hay, staring into the darkness. I kept thinking of the life I had known, and compared it to my present existence and desperate future. I had been given nearly anything I wanted, and I had lived in a country where opportunity was for the asking. This was my first time to lie in total darkness when I wanted light. All my life it had been a simple matter to flick a switch and be surrounded by bright light. I had taken all of these things for granted, but now as I lay in the darkness wishing for a light, I began to realize the value of things I had known. The electric light, like other material possessions, is seldom appreciated until taken away. In the past I thought nothing of a hot bath, clean shave, a fresh pair of pajamas and a soft bed to sleep in. Now I would have given anything to have them. . . .

Steady beams of light from the radiant sun came through my window, changing darkness into daylight. I blinked my eyes focusing them on the window. I had witnessed many a dawn from various places, but this was my first from a prison in Germany.

I heard footsteps nearing my cell and turned as the guard unlocked the door. This time the guard was an old man in his late sixties.

"*Essen, essen,*" he commanded in a shrill voice.

"Toilet," I answered.

"*Jawohl.*"

I followed him to the other end of the hall to a small room half the size of my cell. I hurried in, leaving the old man standing in the hall. In addition to the toilet, there was a wash basin connected to the wall, and a bucket of water on the floor. After I used the toilet, I poured half the bucket of water into the basin and washed my hands and face. The water was cold, clean and refreshing. I dipped my head under the water and washed the cuts on my face. Back in the hall I followed the old man to cell number six. Alston was waiting at the door and greeted me with a smile.

"Ken, did you notice that all of the cell doors are closed this morning? I've already been to the toilet and when I came out in the hall, all of the doors were closed."

"Hell, no, I didn't even think about it. Do you suppose the rest of our crew has arrived?"

"I think several of them are here because the jail is loaded. The old man is going to feed us, so we should know in a few minutes."

We waited patiently while the old guard moved about in the hall opening doors. We heard our own language being spoken and our hopes mounted as we waited in suspense. When the old man reappeared in the doorway, he was accompanied by Gene Coletti and John Mather.

"Gene, you old son of a gun," I shouted.

"Ken! You and Alston! I can hardly believe my eyes," Gene said.

"John Mather, you ignorant Texas bastard," Alston hollered as he squeezed Mather around the waist.

"Hello, you all," Mather said smiling. "They can't kill a damn Texan, can they, Ken?"

"When did you characters get here?" Gene asked, as he borrowed my cigarette and inhaled several drags.

"We got here yesterday afternoon," Alston said. "Gene, what's the matter with your head? You look like a battle casualty with that bandage around it."

"Bob Bulla and I hit the ground side by side," Gene said. "We lit in a farm field just outside of Wiesbaden. The civilians were on top of us when we came down and someone hit me on the head with a knife. Bulla was hit in the head with a sledge hammer and when he fell to the ground a German farmer shot him several times."

"I hope the sons-of-bitches rot in hell. Is Bulla dead?" asked Alston.

Gene gritted his teeth and flinched as he spoke.

"He was killed instantly. His head was split open by that sledge hammer. It was a horrible sight. They shot him, stomped on him, and dragged his body across the field."

I put my arm around Alston as he wept. Bob Bulla had been his pal.

"Well," Gene continued, "it has been like a nightmare to me. I managed to break free of them, and I ran to the highway. I was bleeding from the cut on my head, and I fainted as a German staff car stopped in front of me. They saved my life for when I woke up I was in the local jail at Wiesbaden. A civilian doctor dressed and bandaged my head. Last night Tim Conway and Mather joined me. We had to sleep on a filthy stone floor and didn't get much sleep. Early this morning they took Tim out. He got the hell beat out of him and a bad cut across his face. Mather and I saw him on the public square when they loaded us into a car surrounded by an armed guard. While we were waiting in the car, some German troops carried Wheatly's body by us on a stretcher. They had hung him from the top of a building by the cords of his parachute. We saw him for only a minute. I won't bother to tell you what he looked like. They had let him hang there all night."

"Gene, I guess you know he was dead," I said.

"His neck was broken and twisted," Gene answered.

The guard entered the cell carrying a tray in one hand and a pitcher in the other. Each man was given a cup of coffee, a piece of bread, and a small bowl of soup.

"*Essen*," he commanded as he left us.

Mather still had his pants leg rolled up and I noticed that his leg was turning slightly blue at the edges of the bandages.

I helped Mather with his breakfast and when he finished eating, I readjusted the bandage on Gene's head.

Mather continued the story of their capture.

"I should tell you now that Jim Falla is also dead. We tried to escape together, but he was shot several times by a German soldier who was chasing us. I surrendered to some German guards who rushed us from the other direction."

Gene looked at me.

"That still leaves Terrell Hollis unaccounted for," he said.

I related what the girl interrogator had told me about Terrell.

"I was in doubt about Terrell being dead," I said. "Since you have told me about the others, I am convinced that he was killed too. They even told me Terrell's name and crew position."

Gene set his empty coffee cup on the floor and spoke to all of us.

"We have all gone through a great mental strain and it is still difficult for me to believe that the Germans murdered five members of our crew. Nevertheless, they were murdered in cold blood and the sooner we face it the better. They are dead and we can't help them now, but when the war is over we can do everything in our power to punish the people in this area who killed them. We must always remember Hollis, Poltra, Wheatly, Bulla, and Falla."

Gene pulled a pack of cigarettes from his shirt pocket and passed them around. I supplied the matches and we all enjoyed a pleasant smoke. Gene continued talking.

"We must formulate our plan for future interrogations and review each other on the subject. We should talk our situation over and organize our stories. I feel sure they will take us to Dulag Luft, so we should prepare ourselves for the worst."

While Gene was still talking, the guard returned and took us to our respective cells.

The noon meal was not served. It was late afternoon when the guard came for me. I felt as if days had passed since our early breakfast gathering. We assembled in Norman Alston's cell and had just started eating when a loud noise was heard in the hall. Conway's voice could be heard throughout the building, for he was shouting and cursing at the top of his voice. He was brought in by the indolent, middle-aged sergeant whom I had confronted on the previous day. To our surprise, the sergeant spoke good English.

"Tonight at twelve o'clock, you will leave for Dulag Luft, located at Frankfurt. We will feed you at eleven o'clock. Three of you [Coletti, Mather, Simmons] have no shoes, so you will have to march without them. I advise you to rest, for you will march a long way. You bomb our railroads and you are Luft Gangsters, so you will march everywhere you go in Germany."

Without waiting for questions, he walked away.

I looked at Tim in amazement as he squatted on the floor. His left eye was black. He had a deep gash across his right cheek, and two bumps on his forehead. Beads of sweat dripped from his body. His face was a deep red, and his eyes flashed in anger.

"Those dirty bastards. I'll get even with them if it's the last thing I ever do."

"Take it easy, Tim. What happened to you?" I asked.

"The German Army captured me and several soldiers beat me with clubs. That's how I got this gash on my face. I stayed with Gene and Mather last night in Wiesbaden and early this morning they came after me. A German Major tied me to a motorcycle and dragged me over a gravel road before the civilians. He hit me on the head several times when I called him a cowardly son-of-a-bitch. He brought me here and made me mop the mess hall floor. He poured water on me and hit me in the face with a damn dirty mop. If I get a chance, I'll kill him."

As we ate our meal silently I thought about the sergeant's speech. Dulag Luft. I knew at Dulag Luft we would face the worst test of endurance known to man. The best psychologists in Germany were stationed there, and they seldom failed to get what information they wanted from a prisoner. I tried to remain calm as I spoke.

"Well, fellows, we are going to leave here tonight, so we had better make plans for the future. We have all been told about the practices carried on at Dulag Luft. Let's review everything we can remember that Intelligence gave us in lectures and picture shows. When we get to Dulag Luft we can expect the worst kind of treatment. We know they will bribe us, starve us, and even torture us to get one or all of us to talk. If one of us should break down, you may rest assured that from there on it will be hell for the rest of us. We will be kept in solitary confinement. We will be lonesome, hungry, dirty, and as miserable as they can make us. They will offer us food, cigarettes, and nearly anything to talk. But we must tell them nothing except our name, rank, and serial number. I think it is very important that we refuse to answer any other questions regardless of how insignificant the question may seem. Once they get you beyond name, rank, and serial number they will never stop."

"We can always tell them the wrong answers," Alston said.

"No, no," Gene said. "They would know instantly if you were lying. They already know too much about us. They can look at you and tell if you are telling the truth. You must remember that they have studied people's thoughts and reactions for years. Lying will only give them an excuse for severe punishment. Be polite and to the point, but tell them you cannot answer the question!"

At this point I interrupted because I wanted all of them to swear not to talk.

"We must all realize, as Gene just said, that they are trained psychologists, and are far smarter than any of us. If they weren't getting the job done, they wouldn't be in Dulag Luft. I think it will be plenty rough and we must expect it to be rough. I think we owe it to each other to take an oath here and now. I will start.

"I solemnly swear before Almighty God and you as my witnesses that I will not talk no matter what I may suffer."

Mather looked at me, and repeated:

"I swear."

Tim, Gene and Alston followed.

Gene said:

"You are the best crew a guy ever had, and when the Germans tell me that one of you talked, I will laugh in their faces."

The private of the guard gathered our dishes and returned us to our cells.

I looked out the window and gazed at the gray clouds. I wondered if it would rain. I thought of the long march, and wondered what it would be like without shoes to cover my feet. Walking in stocking feet in the cold October weather had been painful so far, and I knew it would be really difficult on a long march. I sat on my bed of hay, as darkness covered the room, and tried to picture myself in the days ahead. I was much too excited to sleep. Minutes seemed like hours. Finally the long wait was over.

I heard the key turn in the lock, and watched the lights come on as my door opened. The young boy on duty the night before bade me follow him. The clock in the hall said 10:55 P.M. When I arrived at cell number six, I was surprised to see all four of my friends eating their late supper. Few words were spoken and by eleven-fifteen everyone had finished eating, and the conversation became cheerful.

Tim re-enacted his experience. Though it was far from funny, all of us seemed amused. I told them how I had had my face slapped by the German woman, and they accused me of getting fresh. The young guard had returned with a first-aid kit and had given it to Gene. Gene busied himself dressing Mather's leg. I watched with interest and tried to keep Mather from thinking about it. It was nearly 11:45 when a bully appeared in the doorway.

He was a Major, dressed in Luftwaffe uniform and black

boots, and wearing sidearms consisting of a small dagger and Luger. He was about six feet tall, of slender build. Somehow he had lost a part of his left ear. Tim glared at him in horror.

"So, my fat friend is still alive," he said in good English as he looked at Tim. "Let me make myself clear because I am only going to say this one time. I am Major von Capp and I am taking you to Dulag Luft. I lived in New York City for seven years, and I am aware of American trickiness. I consider you a group of gangsters and would not hesitate for one minute to kill all of you, if you make any trouble. You will do exactly as I command if you have any desire to live. Your fat friend has already had a taste of my hatred for air gangsters and I have an accompanying guard of six men, all of whom are excellent marksmen. When we leave here, you will fall into single file, one behind the other. We will march to a small village eighteen kilometers from here, since you have bombed our rail centers in this area. There will be no stops on the way.

"We are due to board a train there at 4:20 A.M. The train will take us to Frankfurt. While on the train and during the entire trip you are to speak to no one. We are moving you during the night in order to protect you from the civilians. Frankfurt was bombed this morning for the fifteenth time by the American Air Force so I have requested an additional escort of twelve soldiers at Frankfurt.

"Let me warn you again that I will not tolerate any disobedience to these orders. If your conduct causes the slightest doubt in my mind, you can expect to be shot. You will not be allowed to smoke when you leave this building. You will address any German officer as sir, and stand at a rigid attention in a German officer's presence. You will lower your head when passing German civilians. They consider you swine and dung and that is exactly what you are. Do I make myself clear, swine?"

The room was silent.

"Very well, fall out immediately and prepare to leave."

As Tim passed through the door, the Major slapped him across the back of the head so roughly that he knocked him to the floor.

"That is just to remind you that you are swine, and I am your superior."

Tim staggered to his feet and followed the rest of us down the hall.

VI

"Luft Gangsters"

OUTSIDE THE SIX GERMAN GUARDS FELL IN THEIR PROPER POSITIONS. One stood at the head of the formation, one at the rear, and two guards took up their positions on each side. At a command from the major we moved off in formation. After two hours of marching, it started to rain and each forward step became more difficult.

Mather, Gene and I were in stocking feet, and our feet were numb. Our socks were, of course, in tatters. Our feet were swollen, cut, and bruised. Gradually the pain crept up our calves until our whole legs throbbed.

As we moved forward our minds were filled with hatred and bitterness towards all Germans. Mather staggered as his injured leg grew worse. He tried to march with the rest of us, for he hated to give the major the satisfaction of thinking he was weak. I kept watching the major. He was a man in his late fifties and I could tell that he was getting weary. I was hoping that Mather could last as long as the major. The major finally stumbled over a rock and ordered us to stop for a rest.

A few minutes later we were ordered to resume the march. At three-thirty a.m. we reached our first destination.

We were given seats on a bench in the village depot. The train pulled in at four o'clock, and we were marched to the far end of the platform. A few angry civilians watched us board the prison car, but no one tried to attack us. I was asleep when the train pulled out. The prison car was small and very uncomfortable, but we were too tired to care.

I was awakened by loud voices in the car. It was daylight.

I looked out the window and was shocked by the sight before me.

We had entered the outskirts of Frankfurt and were moving toward the center of the city. The entire populated area for miles around was leveled to the ground. It looked as if a tornado had swept the city clean of every house, building, and permanent structure. Hundreds of city blocks lay in waste and total ruin. Junk piles of twisted metal, brick, pipe, and broken concrete towered high in every direction. Paved streets, sewer, water, and gas lines had been blown to pieces. Street signs made of reinforced steel were twisted in knots. Telephone poles were uprooted by the hundreds.

As we moved into the heart of the city, vast destruction mounted before our eyes. Black soot and drifting ashes filled the air. We saw several buses and street cars smashed as if they had been made of paper. Lumber, stones, brick, gravel, metal and other debris were scattered about everywhere. Thousands of slave laborers, German civilians of all ages, and German troops were working all over the city clearing the wreckage. The German Army was busy restoring communications and transportation. As we entered the marshalling yards, we saw miles of steel track rolled in twisted circles.

Frankfurt was completely destroyed. It would take several years of hard labor to rebuild this city. Now I was aware of the destructive power of the American Air Force, and I understood why the Germans had such a violent hatred for us.

The train came to a stop. Major von Capp opened the car door and stepped onto the platform. He moved about the crowded depot trying to find his escort. The train platforms, depot, and entire marshalling yard were congested with stranded civilians and soldiers waiting for trains to carry them from the city. They didn't seem to care in which direction the trains were going, so long as they could get out of Frankfurt. Hundreds tried to mob every departing train, fighting to get aboard. I watched the faces passing to and fro, bitter, remorseful, revengeful, hating faces. Cold chills ran up and down my spine, and sweat broke out in my palms.

I wished we had stopped at the edge of Frankfurt and walked around it.

In a few minutes the major returned with an escort of eleven privates and a sergeant of the guard. One of the guards in the car ordered us to fall out. We quickly lined up in single file while the seventeen German guards formed about us. Immediately, women, children, and men of all ages began to congregate in the area. An old man, near at hand, recognized our uniforms and shouted at the top of his voice.

"Luft Gangsters! *Schweine!* Luft Gangsters!"

Many eyes turned in our direction and other bystanders repeated the words:

"Luft Gangsters!"

The crowd continued to grow until the entire platform became so congested that pedestrian traffic came to a standstill. Wild revengeful stares told us that we faced a critical situation.

The major ordered them to clear the way, but no one moved. The old man who had acted as cheer leader for the crowd was eager for a lynching. He pushed one of the German privates aside, raised his walking cane over his head, and struck Gene violently across the head. The crowd cheered his action, as Gene staggered backwards. Major von Capp's hatred was well known to all of us, but he was a soldier of the Prussian type, and he was determined to carry out his orders. He considered it a personal insult and a strong breach of discipline for the civilians to disobey him. He was an officer of the Third German Reich, and he would make them realize that his word was law.

The major jerked his pistol from its holster and pushed the old man back into the crowd. He screamed a command, and the entire guard fixed bayonets and raised their rifles toward the mob. He fired his pistol in the air as a final warning, and the mob stepped back. He gave the order to move off and the crowd stepped back further to give us room. They knew he was dead serious and would not hesitate to fire on them if they refused to obey his command. The way was cleared without further demonstrations.

At the end of the sidewalk a bus was waiting for us. It was painted white with green crosses on its sides; on the sides were the words: Dulag Luft. We were more than anxious to get aboard. As the major took his seat, the bus driver pulled away, leaving the angry crowd behind.

At the city limit we turned onto a small highway leading off to the right. Three miles further the highway came to an end. We were at the front gates of Dulag Luft.

We followed a winding drive to a large brick building inside the camp. I wondered if Dulag Luft would be as gruesome as I had anticipated. Gene, Tim, Alston, and I smiled at each other as the bus came to a stop. We had heard about Dulag Luft ever since we had been in combat, and now: this was it.

VII

In Solitary at Dulag Luft

RECOGNIZED AS THE GREATEST INTERROGATION CENTER IN ALL OF Europe, Dulag Luft had the reputation of being the chief bureau of information for Nazi Germany. Allied Intelligence regarded it as the outstanding military agency for gaining accurate and vitally important information from Allied fliers who became prisoners of war. Intelligence knew that Dulag Luft was a modern, well-equipped prison, operated by an efficient staff of highly trained psychologists.

Each prisoner was studied by several psychologists in order to learn his likes, dislikes, habits, and powers of resistance. The method of procedure was then determined, and the machinery was set into operation to destroy his mental resistance in the shortest possible time. If the prisoner showed signs of fright or appeared nervous, he was threatened with all kinds of torture, some of which were carried out, and he was handled in a rough manner. Others were bribed by luxuries. They were traded clean clothes, good living quarters, food, and cigarettes for answers to certain questions. The gullible ones were told convincing stories of why they must talk and even made to believe that resistance was futile. Those who could neither be swayed nor bribed were treated with respect and handled with care in the interrogator's office, but were made to suffer long, miserable hours of solitary confinement in the prison cells.

Nothing was overlooked by the German interrogators. They studied the results of each interview, and devised new methods to gain the desired information. Dulag Luft had proven so successful that it was without restriction or regulation from even the highest authorities. It was answerable only to the General Staff in Berlin.

Allied Air Corps Intelligence therefore had started a counterattack against Dulag Luft by training every flier in its command on how to act as a prisoner of war. Every method used to gain information from prisoners was illustrated with films and lectures. Interviews between prisoners and their interrogators were cleverly demonstrated to bring out the tactics of the German interrogators. Name, rank, and serial number became the byword of the counterattack. Men were drilled and trained by Intelligence until they knew exactly what to expect and what to do. It became a court-martial offense and an act of treason to reveal any information as a prisoner of war. Patriotism and loyalty were stressed, and American airmen were shown the results of information the Germans had secured from prisoners at Dulag.

By the summer of 1944 the Allied counterattack had begun to show results. My four crew members and I had received the full training course on Dulag Luft, and we entered its grounds with determination to survive the ordeal before us.

As I stepped from the prison bus, I tried to fix a clear picture in my mind of everything worth remembering. The camp was built on level ground and there were no signs of camouflage to be seen. I remembered the large white rocks that covered the length of the front lawn forming the words, "Prisoner of War Camp." The same identification was painted in white letters across the roof of nearly every building we saw. Dulag Luft was of great importance to the Germans, and they knew the Allies would never bomb it as long as it could be identified from the air. We observed new buildings under construction and judged the camp to cover some five hundred acres of land.

The boundaries of the camp were formed by two parallel fences ten feet apart. These fences stood twelve feet high, with trenches and barbed wire entangled between them. There were guards on duty at close intervals from both sides of the barricade. Watch towers were spaced around the camp at one-hundred-yard intervals. Trained dogs prowled the outer boundaries and heavily armed pill boxes were scattered beyond the barbed wire.

We lined up in front of the brick building while Major von

Capp went inside to make his report. Several minutes later, a young, neatly dressed Lieutenant appeared before us. The German sergeant in charge called us to attention and saluted.

"At ease," the Lieutenant commanded in English. "Who is the ranking officer among you?"

Gene weighed the question and raised his hand.

"If you are the ranking officer, then you are the pilot of this crew. You will please call off the name, rank, serial number, and crew position of each man present."

Gene responded innocently.

"I am the ranking officer, but I do not know all of their full names or serial numbers and certainly not their crew position."

"Really, Lieutenant," he said. "Crew position has been added to your name, rank, and serial number, only because your identity must be made certain. Now we will start with you and go down the line."

"My name is Eugene C. Coletti. I am a 1st Lieutenant and my serial number is 0-700721."

"Come, Lieutenant. You haven't finished. How about your crew position? You are wearing the uniform of the United States Army Air Force. You bailed out of a B-24 Liberator, and you were the pilot of that plane. Well, do you deny it?"

"I can't answer the question," Gene replied firmly.

The young officer flashed his green eyes and shouted: "You will answer, Lieutenant. Next."

Each of us followed Gene's procedure, giving our name, rank, and serial number, but nothing more.

"Gentlemen, there are five of you before me today. Since you appear to be good friends, we assume that you are from the same crew. We know that ten men fly on a B-24 Liberator. We also know that ten parachutes opened when your plane was destroyed. That still leaves five of your crew members unaccounted for. Your friends may be held as spies and you could save their lives by giving us their names, rank, serial numbers, crew position, and a physical description of each. If you refuse, we will find out just the same, but you will be made to suffer for resisting us. Do any of you know of their whereabouts?"

We remained silent.

"Very well, we shall see. As you know, this prison is called

Dulag Luft. We are going to identify you one way or the other. As a word of advice, you will find it more convenient to do exactly as you are told and be quick about it. If you refuse to help us you will be punished severely. You will follow me."

We went down the street past several buildings and turned left at the second corner. A block away, we observed the largest building we had seen in the camp. It was "U"-shaped, three stories high and covered two city blocks in area. It was either the main prison or some important administrative building. We walked up the steps to the main entrance on the second floor. Here we entered a long narrow hall with rooms on each side. Two German sergeants in each room interviewed and searched incoming prisoners. As soon as the Lieutenant departed, I was escorted to one of the rooms. The door was locked and I was given a seat. The older sergeant, sitting at a small desk, began questioning me.

"Your name is Kenneth W. Simmons, 1st Lieutenant American Air Corps, and your serial number is 0–717115. Is that correct?"

"Yes."

"Where did you come from, Lieutenant?"

"I came from a small jail outside of Mainz."

"No, Lieutenant, I mean whereabouts in England."

"I can't answer that."

"What is your crew position?"

"I refuse to answer that or any other question of a military nature."

"Very well, Lieutenant. Please remove all of your clothes and stand against the wall."

As I started to unbutton my shirt, the other sergeant jerked me up by the collar and shouted: "You will stand while undressing and you will be quick about it. We will not tolerate inefficiency or disrespect."

I undressed as fast as possible in silence. When I was naked, I handed the sergeant my clothes and stood against the wall. Both sergeants searched my clothes inside and out. When they had finished, I was ordered to raise my hands over my head and bend over. They then proceeded to search every part of my body to make sure that there were no hidden items on my person. The examination was conducted in an abusive manner, and I was happy to be dressed once again. The senior sergeant placed my lighter and two remaining cigarettes in a brown envelope and wrote my name on it.

"Your belongings will be returned when you leave here, that is, if you leave. We are going to start you on solitary confinement since you are pig-headed. We will see how you react. That is all. Take him away."

The young sergeant grabbed my arm and started for the door. As we moved down the hall, I noticed the doors adjoining us were still closed and I wondered if any of my crew members had talked. We made two turns into adjoining halls before reaching the iron gate opening to the prison chamber. The corporal on duty took my card from the sergeant and fumbled through the file box on his desk. He marked the number one hundred twenty-two on the front of the card and placed it in the box. Locking the gate behind us, he escorted me to my cell.

The door was open, but it was pitch dark inside. As I stepped into the darkness, the door closed. I moved my hands along the wall hoping to find a light switch. There were no openings or windows of any kind, and the air was stale. I stood motionless for several minutes until my eyes became adjusted to the darkness. The room was about eight feet square and it was empty except for a small bed of hay and a drinking cup in the corner. I looked around the room and noticed a white sign on the door. "Pull handle for guard." I stepped forward and found a red handle running through a hole in the door. I examined it and moved away to the pile of hay on the floor. Stretching into a comfortable position, I wondered when I would eat again. Fifteen hours had passed since my last meal and my stomach was growling from pangs of hunger. I closed my eyes. . . .

I judged it to be late in the evening when I was awakened by a loud turmoil in the hall. I rose from my bed and listened to loud voices and the rattling of pans in the hall. The noise grew louder. Presently my door came open, and I rushed forward into the light. The guard shoved me back.

"*Nicht, nicht,*" he shouted.

I stood in the doorway amazed at the sight before my eyes. The Germans were serving the evening meal. One guard was pushing a large portable table stacked high with bread down the middle of the hall. Two guards on each side of the table carried large drinking cans. As they reached each cell door, they gave the prisoner two slices of bread and poured him a cup of water. Another guard followed to lock the doors. Most of the prisoners were Americans, and I was shocked by their horrible appearance and disgraceful conduct. Many had been here for some time. They had dirty faces and heavy beards; wild stares, sunken faces, pale expressions were everywhere. They stood with outstretched hands and pleading eyes. They grabbed the bread with fiendish expressions and some begged for more. Several American officers were slobbering at the mouth. The sight turned my stomach and my hunger temporarily vanished. I became frightened, for I knew the prisoners about me were either going crazy, or they were suffering from the effects of starvation. When the guard handed me the two slices of bread he raised the drinking can and shouted.

"*Wasser, wasser.*"

I quickly found the pint cup, and the guard filled it. The follow-up man closed the door, leaving me in darkness. The bread was hard as rock, and I noticed it had streaks of mold through it. I threw both slices on the floor and placed the can of water by my hay bed. I had just witnessed the horrible effects of solitary confinement, and I became aware of the grave danger ahead. I stretched on the pile of hay and wondered if I could survive the discomfort and torture. I prayed to God for the courage to survive. Material possessions, friends, and social position were of no value here. God, if I could reach Him, was the only one who could help now.

Early the next morning I wakened to the clanging sounds of dishes and pans. Hunger pangs were roaming through my stomach as I looked at the moldy bread I had thrown on the floor. When the door opened I looked for members of my crew, but the faces at the doorways were unfamiliar. The German guard filled my drinking can with water and gave me two more slices of moldy bread. I lay on my bed of hay and picked up the two slices of bread I had thrown on the floor. I brushed them off, knowing now I would eat almost anything. I decided to ration my bread and to eat each slice only when I became extremely hungry. I broke one slice into eight small squares and chewed each piece until it melted in my mouth. I spent half an hour eating the piece of bread, and I felt somewhat satisfied when the meal was over. I had saved three slices. I took several sips of water, rose from my bed and moved to the door. I found the red handle and pulled it fully forward. Several minutes passed before the door opened.

"Toilet," I said pleasantly.

"*Jawohl.*"

I followed him into the daylight. The guard waited in the hall while I washed my face and hands and made use of the toilet. I stalled in the bathroom as long as possible, looking out the window at the beautiful scenery. Finally the guard called to me and I returned feeling refreshed and relaxed.

About three o'clock in the afternoon a German sergeant came to my cell.

"Please come with me for a short interview."

I welcomed the interrogation as a respite from the dragging hours of solitary. I hoped it would last a long time, but I was determined to tell him nothing. He started the conversation in a friendly manner.

"How do you like the confinement of solitary?"

"I don't like it."

"Good! I don't blame you, and I think we can end it today. The American Red Cross has given us a form for you to fill out so that they can notify your government and family that you are a prisoner of war in Germany."

He handed me the form and a yellow pencil. At the top were the words, American Red Cross Questionnaire. I had seen this form before during a lecture given by American Intelligence and knew it was a trick to gain information. I filled in the blanks giving my name, rank, and serial number, then crossed out the forty-seven other questions on the sheet. I returned the form to the sergeant.

"Why have you crossed out the other questions?"

"Because I do not wish to betray my country."

"How can you be so foolish, Lieutenant? Question number four asks for your date of birth. Number five calls for your age. Six is your religious preference, and seven is your home address. Do you call answering those questions betraying your country?"

"No, sergeant, I really don't, but why did you stop with seven? Question number eight asks for my group and squadron number. Number nine says to list my missions in the order I flew them. Ten asks for my crew position and question number fifty says to describe the strength of the Allied Air Forces in Europe." I paused to look the sergeant straight in the eye.

"My age and place of birth probably doesn't matter, but I cannot give you any information other than my name, rank, and serial number."

"Do you believe that we made the form up?"

"Yes."

The sergeant glared at me.

"You are a fool and much too suspicious for your own good. Unless we can identify you as a flier, you will be classified as a spy and the penalty is death. Up to now you have been in solitary less than twenty-four hours. Wait until you have been there a week, two weeks, or even a month. Can you picture yourself then? You will have sunken eyes, a hollow face, and you will be screaming for food and water. Solitary in this prison is slow death by degrees. In the end you will either talk, die from madness, or be shot as a spy. It is a slow process, but we have all the time in the world. If you answer the questions now you will be spared this ordeal. You can be released today, and you will be given food and comfort. You will get a bath, shave, and cigarettes. You are smart enough to realize that resistance is futile. No man has ever left here without talking. We will break you sooner or later, so decide for yourself."

"I have decided, sergeant. I am a soldier like yourself, and I am forbidden to tell you more than name, rank, and serial number. I admit that solitary is hell, but you are mistaken on one point. The longer I suffer the more determined I will become to tell you nothing. Each additional hour of hunger and pain will make me more determined. I have much to fight for, and you know I cannot answer your questions."

"It is most regrettable, Lieutenant, that you have made such a foolish decision. . . . I overestimated your intelligence. I know the commandant will be disappointed to learn that you want to resist us. You have just requested more solitary, and I am granting your request. Get out of that chair and follow me."

All afternoon I reviewed the interrogation, and I was positive I had told them nothing. So far, I was very grateful for the lectures given on conduct as a prisoner of war. I had been expecting to see that Red Cross form, and I knew that my worst days were yet to come.

When the evening meal was served, I placed the two new slices of bread in my reserve store and ate the oldest two on hand. I continued the practice of breaking each slice into eight pieces to prolong my meal as much as possible. While preparing my bed, I heard a faint tapping. I listened intently and turned toward the side wall. The tapping grew louder. Then I heard a voice.

"Number one twenty-two, are you awake? This is number one twenty, your next door neighbor."

I answered in a low tone.

"This is number one twenty-two. I am doing fine."

"How long have you been here?" the voice asked.

"This is my second day. How long have you been here?"

"I have been here fifteen days."

"Fifteen days," I shouted. "What are they keeping you for?"

"They found my billfold when I was captured. I had it in my pocket by mistake. I had my group, squadron, European Theater number, and a picture of my plane in it. They have a good start on me and they figure I'll break down sooner or later."

"Where were you flying from?"

"I can't tell you that. That's what the Germans want to know."

"Sorry, I forgot," I replied.

"That's okay. We just can't take chances. They may have microphones hidden in our cells. My name is James Fletcher, and I am from Ohio."

"Glad to know you, Fletcher. I am Ken Simmons, from Texas."

"Another Texan. The guy in your cell before you was from Texas. I never heard anything else for six days. I could tell you the history of Texas backwards."

I laughed.

"I'll try to limit my talk about Texas. How long do they usually keep a man here, Fletcher?"

"That all depends, Ken. Some stay a day, and others stay two or three weeks."

"Is three weeks the longest you know of anyone staying here?" I asked.

"Yes. A guy two doors down in cell 116 stayed twenty-one days, but he has the record so far as I know. The last fellow in your cell stayed six days and the one before him stayed three days. It depends on a number of things, but if anyone on your crew talks they will make it rough on the rest of your crew. Be quiet, here comes the guard."

For the next three days I was not called for questioning. I planned a schedule for survival and worked to carry it out. I spent an hour eating my two slices of bread at each meal, and I made three to four trips to the toilet each day. With each trip to the toilet I stayed as long as possible until the guard called me back to the cell. I spent two hours each day going back to my earliest days of childhood and concentrating on everything I had ever done. I had a recreation period each day, dreaming of a pleasant past experience of my life. Every evening after supper I would count and stack at least five hundred pieces of straw from my bed of hay. Fletcher and I usually talked for an hour or more each night, and I slept twelve out of every twenty-four hours.

At the end of three days of solitary confinement, I had summed up a large portion of my life, seeing each incident in its proper perspective. Seventy-two hours of total darkness and complete loneliness without friends or human necessities can do a great deal for anyone. I made a sad but very important discovery about myself. My values and ideals had been absolutely wrong. . . .

Late in the morning of my sixth day of solitary, two guards appeared at the door of my cell.

"*'Raus, 'raus!*" one of the guards shouted, and I marched away between them. We went out a side entrance, crossed the street, and entered a more modern building built of white brick. On the doors facing the long hall, I read the names of German officers ranking from lieutenants to full colonels. We stopped in the middle of the hall while one of the guards knocked on a heavy oak door. When a bell rang, the guard opened the door and stepped inside. In a few minutes he returned, shoved me inside, and closed the door, leaving me alone with the officer.

The German officer was about forty years of age with a shiny, bald head. He was stout, dignified in appearance, and had keen, penetrating eyes. He rose politely as I came up to his desk.

"Please take a seat, and make yourself comfortable, Lieutenant Simmons." He too spoke good English. "I am Captain Koebeig, your senior interrogating officer. I have called you here for a chat." He reached for a pack of Turkish cigarettes and pitched it to me.

"Have a smoke and relax."

I tried not to act greedily as I slowly drew a cigarette from the pack. I thought of the hours I had yearned for a smoke, and I wanted to grab the package of cigarettes and run, but I deliberately placed the cigarette in my mouth and handed the pack to the Captain. He struck a match and extended it to my cigarette. After several long drags, I said in a quiet tone:

"It has been a long time, as you can see, and I appreciate your kindness."

"Don't mention it, Lieutenant Simmons. I always like to help the prisoners when possible. Unfortunately we are soldiers on opposite sides, and I must perform my duty. From the reports I have received you are a stubborn prisoner and will not cooperate in any way."

"I am not stubborn, Captain. I am just a soldier performing my duty."

"Suppose I were to ask you the number of your group. What would you say?"

"I would tell you that I couldn't answer the question."

"You feel you cannot answer my question, yet your group insignia is clearly marked on the tail of every B-24 in the 8th Air Force. You are a member of the 8th Air Force and as you fly over we decode your group and watch the target you bomb. All I am asking is that you acknowledge what I tell you to be the truth. In this way you won't be telling me anything, and I can establish you as a prisoner of war and set you free."

I returned the Captain's straight glance.

"You seem surprised by what I have just said. One of your crew members has already talked, and the others have acknowledged my questions. Mather, Coletti, Conway, and Alston have all been released. You are the only one left, and I would like to see you on your way. All of them filled out the form that you refused to answer. Do you believe me?"

"No, Captain, I don't believe any of them would do that."

"Well, you are badly mistaken, and I shall prove it by revealing the information they gave about you."

Puzzled by these unexpected words, I yet felt that the Captain could tell me nothing accurate.

"I see that you doubt me, Lieutenant. Well then, you are the bombardier of Coletti's crew. You are a member of the 389th bomb group and your group commander's name is Colonel Miller. You were in the 567th bomb squadron commanded by Major Cane. You were stationed in England seven miles from Norwich and you flew approximately ten combat missions. You carried five-hundred-pound demolition bombs on every mission. You trained as a bombardier in Texas and came overseas in July of this year. Your navigator, Terrell Hollis, was killed by German civilians. He was one of your best friends. Coletti is the pilot, and Conway is the co-pilot. I could continue all day if you wish. Do you deny anything I have said thus far?"

I must have looked amazed, but replied firmly.

"I neither deny nor confirm anything that you may say. The only questions that I am permitted to answer are my name, rank, and serial number. If you think you already know everything, then why ask me?"

"That is my job, Lieutenant. You seem to forget that you are the one being questioned."

He reached across his desk for a picture folder.

"Please look at this for a minute."

I took the folder and turned through it slowly, recognizing pictures and diagrams of the Norden Bombsight. I kept a set expression and returned the folder.

"Don't try to tell me that you don't recognize the Norden Bombsight. You are a lead bombardier and you should know it backwards by now."

"I tell you nothing, as I have already said."

A stern look froze the captain's face.

"I have a proposition to make you. I want you to name five parts on that sight, explain their function to me, and I'll set you free."

"I really don't know one from the other."

"You are a poor liar, Lieutenant. You are a bombardier and you know every part of the bombsight. You are also a lead bombardier and you are an expert at bombing. All of your crew members have confirmed this."

"Captain Koebeig," I said calmly, "you are a good soldier. Even if I did know, do you believe I would betray my country?"

Captain Koebeig grew red in the face and shouted:

"As I have told you before, Lieutenant, I am not being interviewed. Conway and Coletti are both officers of your crew. They have answered all of these questions and more, and they have been released. Do you seriously believe that you could betray anyone by telling me what they have already said?"

"My name is Kenneth W. Simmons. I am a 1st Lieutenant. My serial number is 0–717115. If every member of my crew talks I can tell you nothing more."

"You have just advised me that you are in need of more solitary. We are going to give you the works. You will have plenty of time to think things over. When you change your mind, let me know. The interview is over."

The guard quickly entered the room and executed the "Heil Hitler" salute. Koebeig shouted a command, and I departed with the guard.

Back in the cell I carefully studied the vast amount of information the captain had revealed to me. Everything he told me was correct. I knew that one or more of my crew members had talked. I realized he could have guessed about one or two points. He could have guessed about my being a bombardier and the Germans could have decoded our various group insignia, but no one in that prison except my crew members could have known the rest. I reviewed the rest of his statement. He said I trained in Texas and came overseas in July. He knew the names of my present squadron and group commanders. He knew the number of missions I had flown and my location in England. I didn't need to go over the rest. One of my crew members had told him everything. I began to think of Coletti, Mather, Alston, and Conway. I started reviewing each man's past moments of weakness in an attempt to decide the most likely suspect among the four of them.

I realized that any of them could have had good reasons for their past actions, yet three of them had broken under past crises, and one of them could easily be under dope now. I tried the process of elimination, striking Mather from the list. The more I studied the remaining three suspects, the more suspicious I became of each.

All the next day I kept busy deciding which one might have talked. First, I decided on Gene. Later I changed to Tim and then to Alston. Back and forth I weighed the facts, never reaching a final decision. This was my seventh day in solitary.

Somewhere around eleven that night, Fletcher rapped on the wall.

I had tried to "contact" Fletcher all evening, but no response came from his room. Hearing the signal, I sprang to my feet and crouched alongside the wall.

"Fletcher, where in hell have you been? I have banged on that wall all evening, but never could get you to answer."

"I have been getting ready to leave. My interrogator is setting me free tonight at twelve. He finally gave me up as a hopeless case and told me so this evening. I have been taking a bath and shave and getting my things together."

"I am sure glad for you."

"How did you do today?" he asked.

"I did okay, but one of my crew members talked and told them everything."

"How do you know, Ken?"

I related my interview with Captain Koebeig.

"It sure sounds that way, Ken, but you can't be sure. He might have found out some other way. You must always remember that he is trying to bluff you into giving up. I'll be going in a few minutes, so this is good-by. Maybe when the war is over we will meet again somewhere. Remember what I have told you. Don't believe a thing the Germans tell you."

"Thanks for your advice, Fletcher. I'll take it, though it's hard not to blame somebody."

On my tenth morning of solitary I was awakened by the guards. They escorted me out of the building and we walked three blocks to the east side of the prison, where we entered a single-story white brick building. Captain Koebeig was waiting at the front door.

"Good morning, Lieutenant Simmons. I hate to get you out so early, but duty calls. This is the officers' mess hall. You will follow me."

I followed in silence. We entered the main mess hall and he placed me at the far end of a corner table. He looked at me and smiled.

"You will stand at attention during the entire morning meal. You may watch the German officers eat this fine meal. I want to read you this morning's menu. Let's see—we have fried eggs, chopped meat, fried potatoes, toast, and coffee. I am starving, so if you will excuse me I shall order my breakfast."

I could smell the delicious food as the cooks started to bring it out. I watched them place it on serving trays and I watched German officers select their food. Several minutes passed before my table began to fill up. My stomach ached, and my mouth became dry as I watched them eat. I hated every German alive, especially those officers who left scraps of food on their plates. The scraps looked better than anything I had ever eaten. I became dizzy and light-headed. A beating would have been easier than this. I hated Koebeig's fat guts.

After two hours of this torture, the mess sergeants cleared the trays from the table for the last time, and Captain Koebeig returned.

"Well, Lieutenant Simmons, are you ready to eat some real breakfast?"

I glared at him with all the hatred I could summon.

"You are very stubborn," he laughed. "You have just requested more solitary and you shall get it."

He called the guards, and I was hustled back to my cell. At eleven o'clock I was again taken to the officers' mess hall. This time it was chopped steak, baked potatoes, salad, bread, butter, and milk. Two hours later, when I returned to the cell, I had the dry heaves. From four-thirty to six-thirty I watched them eat supper. When I was taken back to my cell, I collapsed on the pile of hay and didn't awaken until the next morning.

Sometime the next afternoon the guards came again and

took me to another building. I was escorted to the German enlisted men's showers. I stood at the end of the shower room and watched them take hot showers and cover their bodies with soap. I hadn't taken a bath or shaved in two weeks, and my skin itched from the hot steam. An hour later when I was returned to solitary, I was scratching all over.

By the morning of the seventeenth day I had weakened both mentally and physically. I had thought a hundred times of calling the guard and telling the captain everything. Each day grew longer as I weakened with the sickness of solitary. Black circles had formed under my eyes, and my face was chalk white. I could hardly stand on my feet and I blacked out every time I bent over. I had eaten sixty-four slices of bread or about three loaves in seventeen days. I was reverting to the animal stage, and I had used my cell several times for a toilet.

During the seventeen days, I had been interrogated a total of five times. During the last four, the captain ate a full course meal during the interrogation. He reminded me frequently that my buddies were eating good meals, and on three occasions he read the contents of American Red Cross parcels. How I held up I will never understand. I grew hungrier and weaker with each passing day. My resistance sank to the lowest possible level.

Sometime near the hour of darkness, on my seventeenth day at Dulag Luft, a key turned in the door of my cell. There was Captain Koebeig in the doorway. A guard entered the cell with a ladder and climbed to the ceiling to screw in a light bulb. I watched the pleasant glow of the light and rose to my feet as the captain came into the cell with a chair. The guard closed the door and left us alone.

"Well, Lieutenant Simmons, I suppose you are going to continue to hold out," he said.

"Yes, Captain. Now that I have gone this far, I feel sure I can last a long time."

"Cigarette?" Koebeig said, as he pulled a small gold case from his inside breast pocket.

"No thanks," I answered bitterly. "I decided to quit smoking about seventeen days ago."

Koebeig fired up and blew smoke about the room.

"Your cell is rather small. I don't like the odor," he said.

"Yes, Captain Koebeig, my cell is very small," I replied with an air of contempt. "After all, what can I expect from the Germans?"

"That's hardly fair, Lieutenant. You know you are keeping yourself here by resisting us."

"Good," I responded. "Then I'll stay here until I am liberated."

"Liberated?" the Captain chuckled. "Do you believe you will be liberated?" he asked.

"Maybe not today, Captain, but we both know they are coming soon. Each day they get a little closer while your army retreats and moves back. It may be some time, but we both know it is going to happen. Someday I will walk through that door a free man. I hope I am still here because I want to see some of your men get a dose of their own medicine."

"Go ahead, Lieutenant. It sounds interesting. What else will happen?"

"For one thing, I will go back to Texas and live a wonderful life of freedom. I will study and work and accomplish all of the things I have dreamed about in here. Even now, Captain, I have the advantage over you. We both know the inevitable outcome of this war. We both know that by the grace of God I am fighting on the winning side. That is why I can survive this temporary torture. If I were in your place I would probably give up, because there will be no future."

"I am twice your age, Lieutenant Simmons, and one day you will learn that all men like us on both sides suffer in war. I have been through two wars, and I have suffered both times. Is everything really bigger and better in Texas?"

For the next twenty minutes I talked about Texas and Captain Koebeig followed with keen interest. Finally Koebeig interrupted.

"It's getting late, and I have stayed twice as long as I had planned. We work long hours around here, so I must retire. Lieutenant Simmons, I let you talk because you are very loyal. Regardless of our differences, we both respect loyalty and a good soldier. You are leaving here tonight at ten o'clock. That is less than an hour from now."

I tried to retard the sudden joyful shock. I seemed to come alive again. My eyes opened wide, and I tried to hold back the tears.

"I will say *Auf Wiedersehen*. I wish you a pleasant journey."

He stuck out his hand, and I accepted it.

"There is one question I have always wanted to ask you. When Coletti said 'Jump!' and you were in the nose of that bomber, what did you do?"

"I jumped," I answered.

As I spoke I realized that I had made a slip. Captain Koebeig was fully aware that only the first pilot of a B-24 could give the order to bail out. I had clearly confirmed that Gene was the first pilot of my plane.

The Captain banged on the door and handed his chair to the guard. As he stepped into the doorway he turned and smiled.

"You see, Lieutenant Simmons, not even men like yourself are kept here in vain. I have never failed to learn something from a prisoner under my supervision, and I can safely say that the statement still holds true tonight."

He walked down the hall whistling.

An hour later, I had bathed, shaved, and washed my hair. I had been checked out and I stood in the departing room among some fifty American and British prisoners. I felt clean and refreshed, and my spirits were at an all-time high. Actually I was outside of myself. I gazed at those around me and realized I was preparing to leave a hell on earth. The shock was too much to take in all at one time.

I had forgotten that my lighter and two cigarettes had been returned. I extracted one from my pocket and inhaled as a blue flame sprang from my lighter. My throat burned, and I became dizzy on the third drag. I swooned and caught hold of a chair. After several more drags, I laughed out loud. A blurred figure appeared before me and shouted.

"Ken!"

My eyes cleared and Gene embraced me.

"I thought you were gone, Gene," I exclaimed.

"I know, he told me the same old crap about you. Have you seen any of the others?"

"No, Gene, not since the day we came here."

"Well, Ken, I am certain that one of them spilled the beans and we were kept to confirm the story."

"I have the same idea, Gene, and to be perfectly honest, I even suspected you."

"I am glad you did, because he told me you had talked, and I halfway believed him."

"I never answered a question, Gene," I said. "Tonight he asked me what I did when you said, 'Bail out.' Being so excited about leaving, I said, 'I jumped.'"

"Well, Ken, I did about the same thing. He asked me how I liked flying, and I said, 'Just fine.'"

Our conversation was interrupted by a German sergeant.

"Silence," he shouted. "You will line up in double file and answer when your name is called."

By this time many more men had joined our crowd. When the sergeant had finished checking the roll, there were one hundred sixty-eight prisoners in the group; forty-seven English, and the rest American.

Again the sergeant commanded silence.

"You will leave in a few minutes. The accompanying guard will be commanded by Captain Bauch. We will march to the front lawn where you will be given the departure orders. If at any time anyone leaves the formation, you will be shot without question or warning. Follow me."

As the double line passed out the main entrance of the building, twenty-two guards armed with rifles and submachine guns took their positions about us. When we reached the front lawn we were formed into a column of five, thirty-five men deep. The German captain approached our group accompanied by three sergeants armed with submachine guns. The sergeant in command called the formation to attention and saluted. The captain returned the salute and shouted:

"Be at ease. I am Captain Bauch. I am taking you to Weissen for plenty of food and clothing. From Weissen we will go to your permanent prison camp. This guard is here to protect you. If any man tries to break this formation, I have already issued orders to shoot him instantly without warning. I want you delivered safely, so please do not be foolish. I have shot several prisoners, but none has ever escaped from me.

"Because of a possible air raid it will be necessary for the senior Allied officer in this group to sign the parole for the entire group. The officer signing will be American Major Winn. This means in accordance with the Geneva Convention that you have signed your temporary pledge as an officer and that you solemnly swear not to attempt escape while on this trip. In return for this pledge you may take shelter on the ground if we are attacked by planes. If any man should break his signed pledge before we arrive at our destination, he will be shot when he is captured. Even if you should succeed in escaping, your own government is compelled by the Geneva Convention to send you back for military court-martial. To save time, Major Winn signed for all of you. If there is any man who objects, please fall out, and we will return him to a cell in Dulag Luft."

Cursing was loud, but all one hundred sixty-eight men remained in their places.

"Then I may consider the parole agreed upon by every prisoner, since no one fell out. You will not be allowed to smoke while marching or at any time while in the presence of civilians. Ignore any civilian comments, and carry out my orders as quickly as possible. We will board two buses outside the main gate. They will take us directly to a railroad side track where two prison cars are now spotted. It is now 10:45 p.m. We will go aboard as soon as your train pulls in. We will not leave until 3:00 a.m., because I want to prevent an incident with the civilians. Are there any questions?"

The roll was called again, and each man's name was checked from the roster as he boarded his assigned bus.

I took one last look at Dulag Luft. It was a place I hoped never to see again. My last thoughts were of three men who had probably left there two weeks ago. As we moved through the darkness I kept repeating their names: Tim Conway, John Mather, and Norman Alston.

VIII

Red Cross Packages Mean Heaven

TWO MILES EAST OF THE FRANKFURT MARSHALLING YARDS, IN THE early hours of morning, a small civilian train with two prison cars at its end pulled slowly forward. It headed in the direction of the Rhine and its destination was Weissen.

The small train followed the winding, freshly constructed track over rough terrain. It passed into a long strip of dense forest and occasionally crossed a narrow bridge with the rapid waters of an ancient stream below. Slowly, the train moved onward into the open plains of Weissen. The sixty-mile journey took nearly four hours.

Less than a mile this side of Weissen, at a sharp curve, the train stopped and the prisoners were ordered to get out. After both cars had been emptied, the train went on its way. The roll was called in alphabetical order and the prisoners fell into a column of threes facing a long stretch of forest away from the town. The three sergeants carrying submachine guns placed the guards in their proper positions and joined the captain at the head of the column as they moved off.

I felt that I was near the point of starvation. My physical condition had been so weakened that each forward step was an exhausting effort. I also felt that a march of several miles would be impossible for most of the men in our group.

As we moved forward along the narrow country road, the forest grew so dense that full daylight was shaded into semi-darkness. Observing the three sergeants and captain out in front of the column and the unconcerned attitude of the twenty-odd guards about us, it struck me that the opportunity for escape would never be more favorable. By a simultaneous attack the prisoners could have overpowered and disarmed every guard in the column before they had time to unsling their rifles and take up firing positions. The entire escort of twenty-four indifferent Germans could offer no more than minor physical resistance to the one hundred sixty-eight prisoners who outnumbered them seven to one. Without a shot being fired and with the guards as an escort, we would have several hours to make good our escape without suspicions being aroused from the military personnel at the Weissen prison. We were close to the front lines now, but when we left Weissen they were going to move us to a prison deep in Central Germany. If the escape took place now, we would be less than fifty miles from the front lines. By traveling all day and night with the German guard as an escort, we could be across enemy lines by the following night.

As I glanced about I knew that most of the prisoners were thinking of escape. I knew that if one man broke ranks and attacked, the entire group would follow. But deep inside I was sure an attempt to escape would never happen here. This could be the best chance we would ever have, but we had signed our parole. We had traded our word of honor as officers and our countries' word of honor, swearing that we would not attempt escape until the temporary provisions of the parole were over. In exchange we were granted unguarded freedom and shelter in case of an Allied air attack on the formation. Our countries had signed the agreement to protect prisoners of war and each nation had agreed to return a native soldier who broke the pledge of honor and escaped. If a prisoner made good an escape under parole, he would be court-martialed and returned to Germany by his own government. The Germans knew the agreement would be upheld by the United States, so they did not bother to guard carefully prisoners on parole marches.

As I stumbled along, I kept thinking of the wonderful food awaiting us at the Weissen prison. In less than an hour we would see food, probably a real meal out of American Red Cross parcels. I recalled the list of items contained in Red Cross parcels that

Captain Koebeig had tantalizingly read to me many times. I was getting dizzy. I hoped we would go straight to the mess hall without any kind of delay.

Looking ahead of the column, I saw a clearing in the woods. Beyond it there was a military establishment of some twenty buildings. The prison was enclosed in barbed wire with fifteen-foot sentry towers scattered around the outer fence. Sentries were patroling the barriers accompanied by German police dogs.

We passed through the main gate and stopped at the first building. A German major and an American colonel stood on the front steps of the building. The colonel was young for his rank and was wearing an American Air Force B-10 jacket. He talked to the German major for a few minutes and then faced our group.

"Gentlemen, I am Colonel Anderson. I am a fighter pilot in the U. S. Army Air Force. I am here, like you, because I was captured by the Germans. I am the senior Allied prisoner at Weissen, and we run the prison under German supervision.

"Now to the point. You are hungry and you need new clothes. You will be fed first of all. Incidentally, your meal is being prepared now. Later today you will be clothed and given a special American Red Cross kit composed of a sweater, pajamas, hat, gloves, shaving kit, sewing kit, socks, soap, cigarettes, and gum.

"I want you to rest and eat sparingly during your first meal. We have plenty of parcels, and you can have all you want to eat. If you eat too much this morning you will be too sick to eat the rest of the day. You will spend the night here and leave after breakfast tomorrow for Sagan. Sagan is located a hundred miles north of Berlin and it is known as Stalag Luft III. It will be your permanent prisoner of war camp. You will be extremely crowded on the train and you will need a good night's sleep.

"While you are here at Weissen, you must carry out orders promptly and efficiently. A bugle will be sounded for all formations. The sirens will be blown in case of an air raid. In this event you will go to your barracks and stay inside. Are there any questions?

"Good! I hate questions. The corner barracks by the north gate will be your lodgings. We will assign nine men to each room and you can fight for the beds. Please do not bother me with anything. I am busy all of the time, and I'll answer your questions at formation. Keep your mouth shut while inside the barracks. There are microphones everywhere. Don't trust the Germans; they all speak good English. The slave laborers working about this camp are Russian soldiers. We are forbidden to speak to them, so watch your step. If the Germans catch any man giving the Russians food or cigarettes, he will be placed in solitary for three days. That covers everything. Fall out and report to your barracks."

We left the formation and proceeded to the barracks, full of hope. Conversation was loud and morale was high. Gene and I met at the front door and were assigned to room number five. Seven other men were assigned to our room, and introductions were made. We all gathered around a large table in the center of the room while British Flight Lieutenant Henry Rockdale picked up a deck of cards and dealt nine hands of five card showdown. Starting with the highest hand, each man selected his bed. The beds were located against the three walls built in tiers three deep, one on top of the other. I won eighth choice and got one of the top beds near the ceiling.

Promptly at 8:30, while many of us were in the lavatory washing for breakfast, the bugle sounded to the familiar American tune of "soupy."

"Chow," an American voice rang out in the hall.

"Last call for chow."

Those already dressed scrambled out the front door while others rushed from the toilets to join the formation. The two barracks composed of permanent personnel had already eaten, but most of them gathered about our formation, hoping to find a familiar face.

As Gene and I moved off with the formation, a voice from behind us called out.

"Gene!"

Turning to the rear of the formation, both of us recognized and greeted our nose gunner, Norman Alston.

"I came here from Dulag Luft eleven days ago," he said.

"Have you seen Tim or John Mather?" Gene asked.

"Yes," Alston said, as a loud voice commanded: "Group attention. Right face. Forward march."

Norman marched alongside of me until we were in front of the mess hall door. The captain in charge halted us and stepped up on the front porch.

"Gentlemen, your meal is ready. All of us have been through what you are now about to go through. Let me warn you to eat very slowly and do not eat too much. Take it slow at first, and let your stomach start getting back to normal. When you are in the toilet sicker than hell, don't say I didn't warn you. Fall out."

Norman told me he would wait for us on the front porch.

Inside we found twelve plates around each table and assorted dishes of food in the center. The meal consisted of boiled potatoes, barley stew, corned beef, two slices of cheese, bread, butter, coffee, cream, and sugar. There was also half a chocolate bar, a package of gum, and a pack of cigarettes by each plate.

We stood, twelve men at a table, while the captain prayed.

"Almighty God, we thank Thee for these fine brave men who have come from the gates of hell at Dulag Luft. We thank Thee for this fine food delivered into Germany by the American Red Cross. We thank Thee for the safe journey of the food trucks from Switzerland. We give special thanks to Thee for sparing our lives. Bless these young men and give them the courage, love, and hope to survive the trying days ahead. Amen."

When the meal was concluded some were already sick, and were dashing from the mess hall.

Outside, Gene and I joined Norman on the front porch and fired up a wonderful cigarette.

"Why have you been here so damn long?" I asked Norman.

"I am a member of the permanent party serving on the entertainment staff for incoming prisoners. When prisoners stay more than one night we give them a stage show. I sing and play the banjo. I asked for the job, and it sure beats going to central Germany," he said.

Gene interrupted Norman with a direct question.

"Where and when did you see John Mather and Tim Conway last?"

"It's a long story, but I have been wanting to tell both of you. On our second day at Dulag Luft, Mather and I were released together. We were interrogated early that morning but neither of us told them a thing. We also refused to fill out the Red Cross form. We could only figure it was because we were enlisted men. We came here with a train load of enlisted men and three British officers. The second day we were here Tim came in with a group of Royal Air Force boys and two American paratroopers. We naturally questioned him about you two not being with him, and he was reluctant to discuss the matter. He and Mather stayed here three days and when neither of you showed up we started questioning him. On the morning of the third day before they were to leave, we really started to work him over. I asked him to explain how it was possible for him to be released while both of you were still being held. Mather kept talking about the three officers of our crew and why one would be released while the other two were held. He became extremely nervous. The third morning after breakfast, Mather followed him to the barracks and accused him point blank of betraying the crew. He told us nothing, and both of us were still nonplussed."

"What did Tim say?" Gene shouted.

"He just looked at Mather. I guess he couldn't say anything."

"Don't jump to conclusions," I said to Gene. "Hell, I suffered as much as you did, and besides Tim probably stood all he could take. We'll see him before it's over and find out just what did happen." And then I repeated what Fletcher had told me about the Germans having many ways of getting information, so that one couldn't be sure that someone had talked. Gene agreed that the case was a mystery.

Without allowing Gene to continue, I started telling Norman about our long stay in solitary and the many different interrogations we went through. We were still talking when the bugle sounded.

"That's the clothing formation," Norman told us. "You are going to be clothed and issued American Red Cross kits to take with you. They will also search you, take your picture, and make you a prison card. Tomorrow they will notify your closest of kin that you are a P.O.W. When you get back with a bath and new clothes you will feel like a million dollars. I'll come over this evening for a visit."

During the rest of the morning we were taken to three different buildings. In the first building each man was interviewed by a

German officer and given a Red Cross form to sign, addressed to the A.R.C. in Switzerland. The card stated that we were prisoners of war in Germany and were being treated very well. In the second building we were fingerprinted, photographed, and given the A.R.C. special clothing kit. The third building was a small clothing warehouse, and the entire group was issued shoes and overcoats. These also were donated by the A.R.C. Was I grateful to get shoes at last to protect my sore and swollen feet!

Immediately following lunch, Gene and I returned to the barracks and prepared to take a shower. We stayed in the water for nearly thirty minutes, and then took turns shaving each other as there were no mirrors in the lavatory. We laid out new socks and underwear and dressed among the crowded occupants of the room.

That afternoon Norman visited us in the barracks and we took another walk around the grounds. At supper we ate the evening meal together. We stayed in the mess hall better than two hours listening to a G.I. who entertained us at the piano with familiar American music, and we talked about each member of the crew who had been killed.

Back in our room most of the fellows were playing cards and talking about England and the United States. I hurriedly undressed, bid Gene pleasant dreams, climbed up to my third-story bunk, and sank deep into the soft hay.

Early the next morning I woke up listening to a bugle. Reveille was sounded promptly at five and by six-thirty our group had packed, eaten their breakfast, and were standing in the assembly area waiting for departure. Our farewell with Norman was very sad. Each of us was dressed in an overcoat, and we carried our clothing kits under our arms. The same escort of twenty-four German guards that accompanied us from Dulag Luft took their places in the formation. As we passed through the last barrier, and entered the woods on the small dirt road to town, I took one last look at Norman and waved. He waved back.

IX

The Worst Railroad Journey of My Life

WE BOARDED TWO PRISON CARS CONNECTED TO THE REAR OF A passenger train. Eighty-four prisoners and two guards were assigned to each car, and the doors were locked from the outside. Both cars were small, only half the size of an American chair car. There were six separate compartments inside the car. Each compartment had two built-in benches capable of seating three persons to the bench. Above the benches were luggage racks, and the bottom floor space was large enough for two men to stretch out on.

Major Winn assigned ten men to each of the six compartments, placing six men on the benches, two on the luggage racks, and two on the floor. The other twenty-four men had to stand in the narrow corridor adjoining the compartments, as there wasn't enough room to lie down or even sit down. Then a sixteen-hour schedule was set up whereby each man would spend four hours on the floor, rotating to the bench, then to the luggage rack, and finally end up by standing four hours in the corridor. The schedule was repeated every sixteen hours until the trip was over. The two guards had a bunk at each end of the car; they were bribed with cigarettes into sharing their quarters with prisoners who became critically sick.

After quarters had been assigned, Major Winn took a position in the center of our car.

"At ease. Captain Bauch, the German officer in charge, has

taken up quarters in a pullman several cars ahead and he has given me some orders that must be carried out. We are reminded that we are still under parole and have pledged our honor not to attempt escape until the parole is up when we arrive at Sagan. We cannot smoke while in any German town or in the presence of civilians or Nazi youth organizations. Food will also be kept out of sight, and there will be no eating in the presence of civilians at railroad stations. This trip will be plenty rough, and you may think it looks bad. Remember, it could be a lot worse, and it will be if you don't carry out my orders to the letter.

"The trip will last three days and three nights. You will have good reason to bite your neighbor's head off, but you must be tolerant and understanding. You will lose sleep, you will be waked up to change shifts, you will be stepped on, pushed, and shoved. Remember that the fellow next to you didn't do it intentionally and forget it. We have a full food parcel for every man here. Actually there is more food than we can eat. The Germans have given us three slices of bread per day, two pieces of cheese and a piece of sausage. Eat all you want, but don't waste anything. The toilet is in the far end of the car. Keep the car clean, be considerate of all others, and do nothing to antagonize the civilians. Carry on."

By trading about, Gene and I managed to get together in compartment number two. I drew the standing shift while Gene was more fortunate in drawing the floor shift. The train had been moving more than an hour before we took up our positions. Each man tried to cooperate with his neighbor, so as to make the trip as pleasant as possible.

I stood by a window covered with cross bars and watched the scenery as the train moved slowly forward. I felt sure it would take seven or eight days to make the trip at the slow pace we were moving. I wondered if we would encounter an air raid, and if so, would the Germans keep their promise and let us take shelter.

We were going to Sagan, halfway between Berlin and Breslau, Germany. This was to be our permanent P.O.W. camp and I wondered what it would be like. I knew there would be many American prisoners there, because it was one of the largest in Germany. It was probably composed of officers since there were no enlisted men aboard our train. It must be composed of flying officers because I noticed for the first time that all of our group were either members of the American Air Force or the Royal Air Force.

After a while the train stopped in a small German town. I guessed we would be staying for a while, when I saw the guards leaving the car to fill the water cans. Many men had started eating and others were fixing sandwiches. The civilians did not notice us. Gene fixed me a sausage and cheese sandwich and when the water came he mixed a tin can of powdered milk. He also passed me a piece of chocolate.

I had just finished lunch when the major signalled that it was time for the second shift. Gripes and varied complaints were now heard throughout the car as the tired prisoners slowly changed places.

Each hour seemed to creep by as the group became more uncomfortable. Every man tried to hold his temper, but patience slowly became exhausted. The eighty-six occupants of the small car were jammed into every available inch of space. There was no privacy for light sleepers or suitable space for those who wanted to stretch. Some men talked while others tried to sleep. Those on the floor had nothing to soften the hardness; occasionally a foot stepped on a back or hurt a finger. When one man screamed with pain, he would wake everybody. Those on the luggage racks had to lie on their sides; if they turned in any direction they were almost certain to fall on the floor. Sometimes a man in the luggage racks would fall asleep and come tumbling down on those on the floor. The hard benches were like sitting on bricks. Four hours of standing was even more tiresome. Obnoxious personalities were forced upon those who ordinarily would have walked away from them.

During the night the situation grew steadily worse. Total blackout on the train increased the discomfort. Clumsy prowlers and visitors to the toilet stepped on everything but the floor. Some quietly relieved themselves on the walls and floors rather than risk stepping on men while going to the toilet. The odor became terrible and everybody started raising hell about it.

The next morning Major Winn was sick from the odor and made another speech.

"At ease," he shouted. "We are never going to make it to Sagan unless all of you carry out my orders. The next man who takes a leak on the floor of this car is going to get the hell beat out of him. You are to use that toilet if you step on twenty men getting there. We are going to use all of our belts to tie those who are in the luggage racks so they won't be falling on other men. From now on if a man gets sick and has to puke you will call out 'Clear the way.' That is the only time 'Clear the way' will be used and you had better clear the way. I am getting damn sick and tired of this bitching. We are going to organize singing groups, and I want everybody to sing. We will sing several hours per day, according to the schedule Captain Jones has prepared. When we are not singing you can play cards, talk, look out the window, or just keep your damn mouth shut. No more bitching, do you all understand."

The sing-song started two hours later. We sang every song known to man. Captain Jones organized a quartet and they sang between times. One fellow went around quoting poetry and one read to us from the Bible twice a day. All of the new activities, added to the interesting daylight scenery, helped to relax nerves and strengthen morale. The songs prevailed on into the night and humorous jokes were told and passed from compartment to compartment. Men grew tired and weary, but all of the activity helped curb their discomforts. At the end of the second day it became car policy that any man who was stepped on, fallen on, or got a mashed finger would look at the offender and say: "Dear Lord, forgive him, for he knows not what he is doing." I heard that a hundred thousand times on the rest of the trip.

It was around midnight on the second night when voices began to spread swiftly about the car.

"Leipzig. They are sidetracking us. We are being disconnected from the train. Leipzig. Leipzig."

As the word spread around, men rose from all positions to get a look. Signs of recent bombings could be seen everywhere. Leipzig was a regular air force target. Fear mounted from the long delay in the marshalling yard. All of us looked momentarily for an attack from the air. Finally, in the early hours of morning the prison cars were connected to a new train and we moved forward once again in our long journey.

On the third day, the singing continued. The same jokes were told for the fifth time.

Gene and I finally got in the compartment at the same time and fixed our lunch together.

"I am beginning to think that those who died are a lot better off. If it is going to be like this from now on I would rather be dead," he said.

For lunch we had spam, bread, cheese, and powdered milk. After lunch the singing started again. I was getting sick of that damn quartet myself, but I pretended that it was delightful.

During the last night of the trip we prisoners were so exhausted that conversation was mild and nearly everyone slept. We were too tired to talk or sing. We were too tired to stay awake, and though each man was more than ready to jump down his neighbor's throat, he was too worn out to make the effort.

During the early hours of the next morning, the train pulled into Sagan and the cars were sidetracked. It was several hours past daylight when the guards finally arrived, and we were taken off the train.

I was dead tired as the new guard formed about us. The roll was called and checked. Many men were sick and could hardly walk. All of us were sore and stiff from the long journey in cramped quarters.

As we marched to the outer boundaries of the camp, my eyes roved in every direction, and my mind was wild with excitement. All of our guards and the sentries about the camp were dressed in Luftwaffe uniforms. As far as I could see there were acres upon acres of buildings and barbed fences. This was to be our new home until the end of the war. I knew that a strange new life awaited us.

X

Initiation into Stalag Luft III

STALAG LUFT III SPREAD ACROSS SEVERAL MILES OF FOREST LANDS outside of Sagan, which was located halfway between Berlin and Breslau, Germany. By November, 1944, this prison for American and British airmen held ten thousand officers in its old compounds, and some five thousand others in new adjacent compounds under construction. It held more American prisoners of war than any prison camp in Europe.

The main area of Stalag Luft III was composed of five compounds known as North, South, East, West and Center. Each compound was about one-fifth of a mile square enclosed by two twelve-foot barbed-wire fences. These fences were parallel and six feet apart. The space between was filled with tangled barbed wire. About twenty-five feet inside the barrier, running parallel with the fences, there was a small steel wire supported by stakes two and one-half feet above the ground. This wire was known as the warning wire; any prisoner crossing it was shot without warning. Immediately outside the fences were sentry towers, fifteen feet high and spaced a hundred yards apart. Each tower was equipped with a searchlight, siren, machine guns, and trigger-happy sentries.

Guards were stationed between the towers. They were armed with rifles and pistols, and had trained German police dogs. The guard was maintained twenty-four hours a day. It was increased during darkness, and doubled in case of a night air raid or other emergency. German administrative buildings were scattered throughout the main camp area, while barracks for German personnel were located just outside the five compound barricades. The entire camp was surrounded by dense woods.

The prisoners were divided equally among the compounds with each compound housing about two thousand men. The British and the Americans were separated. In each compound there were ten barracks, a hundred feet long and twenty-five feet wide. Each barracks was built to house one hundred men, but housing shortages made it necessary for two hundred men to live in each hut. There was an outside latrine for every two barracks and each latrine provided twenty seats and six urinals. Some four hundred men of two barracks shared the twenty seats in each latrine, and this caused continuous latrine lines throughout the day.

In addition to our quarters, there were two small mess buildings, used for issuing German rations, storing food, and serving hot water four times a day. The other two buildings in each compound were a library and a large auditorium. The library was stocked with thousands of American Red Cross books. The auditorium was used as an assembly hall, church, and theater, all in one. Between the outer buildings and the warning wire, there was a path fifteen feet wide in the form of a race track. It was known as the perimeter track and was used exclusively for walks for the sake of exercise and talking to friends.

The space between each barracks was known as the outside laundry. There were many wash stands where washing was conducted daily by each combine. There were fifteen clotheslines per barracks for clothes to dry on. The remainder of the open ground, located in the center of the compound, was called the parade ground. This large space was used for compound formations and games. Each compound was run and managed under the direction of the Senior Allied Officer and his staff.

On the morning of November 6, 1944, my companions and I entered the separation center of Stalag Luft III. The process of separation lasted several hours. Each prisoner was searched, issued prison tags, fingerprinted, photographed, and interviewed.

The British and Americans were separated into two groups and taken to a large bath house for showers. I was amazed by the stingy attitude of the Germans toward water. After all of us had undressed, we were taken in mass formation into the large shower room. There were twenty-four water spouts dependent from the ceiling. One hundred twenty-one men in my group crowded under them. A German sergeant standing at the far end of the room commanded our attention in English.

"Gentlemen, you are all new Kriegies. I will explain the rules for taking showers. You will enter the dressing room at the same time, and when all men have undressed you will assemble in this room. There are only twenty-four showers, so it will be necessary for five men to share each spout. The hot water will be turned on for five minutes, but not one second more. Following the five minutes of hot water, you will be given two minutes of cold water. Share the showers among you, and be very fast. You are given only one bath a week. When you have finished drying off, you will dress and fall into formation outside. Are you ready?"

He raised his arm to the corporal in the corner and looked at his watch. As the second hand reached zero, he dropped his hand and the water came on. He called off each minute of time as we hurried to finish our baths. Men were pushing and shoving when the hot water went off. Most of us were covered with soap, and many others were still waiting to get wet. Two minutes later, when the cold water went off, everybody was raising hell. Half of us had to dry with soap all over us.

All of the men were bitching when the sergeant commanded attention outside.

"You didn't do so well today because you were confused. Do not be discouraged, gentlemen. Next week when you visit us you will be better organized and you will learn to bathe like old Kriegies. You have been assigned to the center compound. Follow me."

I wondered what the sergeant meant by calling us "Kriegies." I nudged Gene and asked:

"What did he call us?"

"Hell, I don't know, craties or something like that."

The Center Compound had been notified of our arrival, and over a thousand American prisoners had gathered around the main gate and theater building. A German lieutenant stood on the outside and an American major on the inside, each counting us as we passed through the gate. When the one hundred twenty-first prisoner passed through, the major went to the head of the column and told us to follow him. The old prisoners cleared a path to the theater. They eyed us closely, but spoke to none of us.

When we were seated inside, a tall American colonel stepped through the front door, and the major called us to attention. He executed a snappy salute and the colonel returned it as he climbed the steps leading to the stage. Our group remained silently at attention. I was surprised by the colonel's appearance. He was dressed in forest green pants and an officer's wool shirt. His shoes were polished, his uniform was clean and neatly pressed, and he wore the wings of a command pilot. He appeared to be a man in his early forties. His head was bald, he had blue eyes, a rugged face, and a pleasant smile.

"At ease," he commanded in a dignified manner.

"I am Colonel John Spivey, Commanding Officer of the Center Compound. The next few days will be somewhat confusing for you new men. That has been the case with me and every older man who came through those prison gates before you. Being prisoners, we are forced through circumstances to do things we had never dreamed or conceived of doing before. We are called upon every single day to sacrifice, endure, and work together in the common struggle for our survival.

"The prisoners of this camp have organized the compounds like any organization in the American Army. Discipline here at Sagan is the most rigid discipline you will ever know in your life. If you break even the most minor rule you will pay dearly. You will learn that we can't afford to make even minor mistakes. American Brigadier General Charles Vandermann is in command of the camp. I command your compound. My staff supervises its operation. Every barracks is divided into two blocks. Each block has a commanding officer. Every block is divided into a number of combines and each combine has a combine leader. We have learned that through organization we live better, eat better, get along better, and in general, life is much more bearable for all of us."

The Colonel paused at this point and glanced about the open windows and doors, looking over the old prisoners who had gathered to hear him.

"I call upon you to do your job to the best of your ability, and to do it good-naturedly. Listen to the older men for they can help you. Treat them with respect for they have earned it. Some of the older prisoners may bite your head off. I ask you to try to hold your temper. Some of these men have been prisoners at Sagan for over twenty months. Their nerves are strained; they are sometimes easily antagonized, and you must try to understand. You will find that the older prisoners have gained judgement and wisdom.

"Brotherhood, kindness, and understanding will take you a long way while you are here. Don't be a slacker, for you will find yourself without a single friend. Life will be only as pleasant as you make it for yourself."

The colonel pulled a cigarette from his pocket and asked us to join him in a smoke. He inhaled two short drags and lowered the cigarette.

"We do not know who you are. We know that we have one hundred twenty-one new men before us who are supposed to be Americans. We intend to make sure you are—to the last man."

Surprised faces and displeased eyes greeted this statement.

"At ease, gentlemen," the colonel continued.

"You are probably Americans to the last man, but we have had German stooges in these groups before who speak perfect English and have every characteristic of being American. There are many things, which you will soon learn about, which the Germans would like to know. These men live among us as spies for the Germans. We have eliminated them to the last man. We do not intend to allow another stooge in this camp.

"This identification will take only a few hours, but until you have been identified as an American, the old prisoners will not speak to you. You will be asked many questions, and you may be placed on exhibition so that someone can identify you. You must be identified by our Intelligence before any of us can accept you. The sooner this is over, the better. Major Thompson will now take over. We will have these talks from time to time. Thank you for your attention."

Major Thompson called the group to attention as the colonel departed through the front door.

"At ease," he said. "I am Major Thompson, Assistant Intelligence Officer. I am going to send you to the library for interviews with my staff. When each of you is identified you will be assigned to your barracks. If we have not identified you by noon you will remain in the library for the noon meal. We will feed all of you the meal; but starting tonight you will eat in your combine, unless you still have not been identified. If we have not identified you by five o'clock, you will be assigned to a special isolation ward until you are cleared with Intelligence. The six lieutenants at this desk in front of the auditorium will take your name, rank, and serial number. You may talk until all names have been recorded, and then we will go to the library."

He paused and then continued: "Just one more thing. The next two or three days, I will be sending men from our Intelligence to interview you. They will ask you questions about the war, its progress, your duties in the Air Force, news about the U.S., Dulag Luft, your trip here, and many other questions. You will answer any question they ask you, for this is a part of our system of information. O.K., give your name to any one of these six officers."

We were astounded at such a complex organization in a prison camp. We formed six lines leading to the desk. When all names had been taken we were formed into a column of fours and marched to the library. A sign on the door read: "Closed all day for identification of new Kriegies."

"There is that word again," Gene mentioned to me.

"Yeah, I think I know what it means now," I answered. "I think it is the German word for prisoner of war. I suppose that is what they call us."

We found seats in the two large reading rooms and were conversing loudly when Major Thompson came in.

"At ease," he shouted. "In the room across the hall I have ten teams composed of three men each that will do the interviewing. We will take ten men to start it off, and as each man is cleared another will be called. When your name is called, go immediately into the room to save time."

The processing started. By noon ninety men from my group had been called, and eighty-seven had been cleared and assigned to blocks. The three remaining men were returned to our group for further examination. There was a constant stream of old prisoners coming to and from the library to check facts and figures and to aid in establishing the new prisoners' identity.

Lunch was served in the library in cafeteria style. Each of us was given a cup of coffee, two slices of thin German bread, and a bowl of thick potato soup. The interrogators ate with us, and by one o'clock the processing started again. It was nearly two o'clock when Gene and I were called in with the second group.

I was given a table by the door, and Gene was ushered to the opposite end of the room. The senior officer at my table was a captain. He sat at one end of the table and seated me at the other end. The other two interrogators sat across from me with paper and pencil. The captain smiled.

"Glad to welcome you to Sagan," he said. "I am Captain Sam McGee. These two gentlemen are Lieutenants John Rhyne and David Abernathy."

We shook hands. When we had taken our seats, McGee eyed me with suspicion.

"You are an American citizen?"

"Yes."

"What was your crew position?"

"Bombardier."

"What group and squadron did you belong to?"

"The 389th Group and the 567th Squadron."

"What Air Force?" Abernathy asked.

"Eighth."

"Where were you stationed?" Rhyne asked.

"Hethel, England, seven miles from Norwich, England."

The three officers took turns asking questions as Rhyne and Abernathy recorded every answer. McGee continued:

"Where are you from?"

"Texas."

"What part of Texas?"

"Dallas."

"What was your last street address in Dallas?"

"714 South Winnetka in Oak Cliff."

"What is Oak Cliff?"

"It is a part of Dallas."

"What high school did you attend?"

"Sunset High."

"Where and when were you born?" Abernathy questioned.

"Lake View, Iowa, on March 13, 1921."

"Have you seen anyone in the camp who may know you?" Rhyne asked.

"No."

"Did any of your crew members come here with you?"

"My pilot, First Lieutenant Eugene Coletti, came with me and my co-pilot, Tim Conway, should have come about a week ago."

"When did you come overseas to England?" Rhyne asked.

"On June 4, 1944."

Rhyne looked at me in a questioning manner.

"Name me three bombardier fields in Texas?"

I told them I would have to think for a minute.

"Let's see. Big Springs, San Angelo, and Midland, Texas."

"Which one did you train at?" Abernathy chimed in.

"I received my bombardier training at Big Springs from October, 1943, until March, 1944."

"Can a B-24 fly without a pilot or will it fly by itself?"

"It will fly by itself with the automatic pilot, but the pilot is needed to operate it."

"What sight did you use and what is its most important mechanism?"

"The Norden Bombsight, and its most important mechanism is the rate end."

"Where is Eugene Coletti from?" McGee said.

"Brooklyn, New York."

"Is he single or married?"

"Married."

"What is his wife's name?"

"Betsy."

"When did you fly your first mission?"

"Around the middle of June."

"What day in June?"

"Hell, I can't remember, the 15th, 16th or 17th," I answered.

McGee nodded to the others as they left the room to check my statement, and to call witnesses to check further my identity.

"Simmons, what was your group commander's name?"

"Colonel Claud Miller."

"What was the name of the Catholic chaplain on your base?"

"Father Beck."
"What was the name of your mess officer?"
"I never met him. I don't recall his name."
"Did you go to college?"
"Yes."
"Where?" asked Captain McGee.
"The University of Texas."
"Did you belong to a fraternity at Texas University?"
"Yes, I am a member of the Phi Kappa Sigma."
"Texas University is in Dallas, I believe you said."
"No, I said I lived in Dallas. The University is in Austin."

The questioning continued for another twenty minutes, and McGee wrote the answers on three additional sheets of paper.

When Rhyne and Abernathy returned, they had another old prisoner with them.

"Lieutenant Simmons, this is Lieutenant Bill Johnson from Dallas, Texas."

We shook hands, and Johnson sat across from me.

"Lieutenant Simmons, on what street is Sunset High School located?"

"Jefferson Avenue."
"What street has all of the downtown theaters?"
"Main Street."
"What are the two main hotels in Dallas?"
"The Adolphus and Baker."
"What is the name of the Dallas Charity Hospital?"
"Parkland Hospital."
"Name the two leading newspapers in Dallas."
"The *Dallas Morning News* and the *Dallas Times Herald.*"
"Who is the present Mayor of Dallas?"
"Hell, I don't know."
"Who is Dallas County's present Congressman?"
"Wilson."

Johnson turned to McGee.
"He checks out perfect."

McGee looked at Abernathy and Rhyne.

"Everything checks. We have talked with five men to check his story. We also checked the information about his pilot," Rhyne said.

"O.K., Ken, you are now one of us," McGee said. "You are

assigned to block seven located in the east end of barracks 152. I am the block commander of seven. I know you will like our group. Welcome aboard and report to Lieutenant Ferrell. He is my assistant block commander. He will take it from there."

When I walked out the front door, a runner asked me my name and block, and I followed him across the parade grounds. We entered the east end of barracks number 152. Lieutenant Ferrell was waiting in the first room.

"Simmons," he said as we shook hands, "how do you like Sagan by now?"

"Fine, but that interview made me feel like a damn German."

"I know, but we cannot take the slightest chance. You will begin to understand tonight. Your pilot, Lieutenant Coletti, has been assigned here and I thought of putting both of you in the same combine."

"Swell," I said. "That's nice of you. Say, what is a combine and what does Kriegie mean?"

"Well, this barracks is divided in the center into what we call two blocks. Each block has one hundred men and it is divided into combines of twelve men each. The combine is where you eat, sleep, and work. The combine is your home. The men in your combine will be more closely associated with you than anyone else.

"The word Kriegie was made up by the German goon guards. It is the name they give to American prisoners of war. They call us 'Kriegies' and we call them 'goons.' We named them goons because of their stupidity, but never call one of them a goon to his face."

The conversation was interrupted by a young Kriegie who appeared in the door.

"Come in, Clark," Ferrell said. "Kenneth Simmons, this is Byron Clark, First Lieutenant, ex-B-17 pilot, and your combine leader."

I extended my hand as the blue-eyed Lieutenant approached me.

"Clark is a fine leader, and I know you two will be great friends. The combines are lettered in alphabetical order. You are in combine 'C'. The combines are lettered 'A' through 'G'. Clark will introduce you to the guys in your combine."

INITIATION INTO STALAG LUFT III

"I will show you around," Clark said.

I followed him down the corridor.

"This is the kitchen. If you are wondering how a hundred men cook here, I can't blame you. We have a mess officer who rations our food and carries out the cooking schedule. We cook by combines. Each combine is given an equal amount of time on the stove. We change schedules every week; thus we eat at different hours. Each combine has thirty minutes on the stove to cook its meal. The meal is prepared by our combine cook and K.P.'s. The cook roster changes every week, and the K.P. roster changes every day. All of us cook and serve as K.P.'s. All of our food is pooled and we eat whatever the cook fixes for the meal. The cook is given our entire weekly rations, and he is responsible for planning our meals. A cook has to be very thorough in planning in order that the food will last for the entire week. We always keep a week's rations saved in case the Red Cross trucks are late, or in case we are cut off."

I looked inside the kitchen. There was a large stove by the window. It was capable of supporting four vessels at a time, and it had a small oven attached to its side. The stove was fueled by a pile of coal in the corner of the room.

We moved along the hall past the various combines. The block was divided into eight separate partitions, four on each side of the hall. Each combine was small, being twelve feet square. Many men eyed me when we passed by. The partitions had been formed by triple stacks of beds that rose from the floor nearly to the ceiling. There were four closets, facing the hall, two on each side, which formed an open passageway to the combine.

We entered combine "C." Clark stepped forward and spoke.

"Gentlemen, this is one of our new members, Kenneth Simmons. Ken, meet the combine."

Clark then proceeded to introduce each man, and I eagerly shook hands as the introductions were made. I was amazed by their neat appearance, cleanliness, and good physical condition. Nearly all of the men at Sagan looked better than I had expected. When the introductions were completed, Clark commanded attention.

"With the addition of Simmons and his pilot, Coletti, we will have a full combine of twelve men. We have been instructed to help these men get situated. I want them to have a good meal and go to bed. They had a hell of a journey, and are in bad need of sleep. Keep the questions for tomorrow. Tom, will you get Simmons' pally-ass, blanket, and pillow, and I will arrange his closet."

When we were in the auditorium that morning, each of us had marked his name on his clothing kit and left it in the auditorium. Clark had sent for my kit as soon as I was cleared by Intelligence.

"Ken, your clothes are in the bottom of this closet. One of our men got them right after you came to the barracks. Here, let me show you."

Tom Pirtle left the room to gather my bedding and fix my pally-ass. That was when I learned the Kriegie name for mattress. Clark opened the door to closet number four and arranged my clothes. Each closet was seven feet high and two feet wide.

Clark looked at me and started explaining again.

"Each closet is shared by three men. These shelves are known as top, middle, and bottom. Each man has a shelf according to his seniority at Sagan. Since you are the newest in the group you will take the bottom shelf and work up. We don't have room to hang clothes in the closets. If you hang any clothes outside, they must be hung at the foot of your bed. Your bed will be either of these two top bunks. Two men moved out of the combine, so we reassigned beds before you came. Again seniority has its choice, and we work down on beds. The younger men sleep on the top and the older men sleep on the bottom. Lie down now on my bed and get some rest. We will eat at seven tonight, and you look tired as hell. Rhinehart, you had better start dinner."

"Hell, I'm ready. You K.P.'s get out of the sack!"

I stretched out on Clark's bed and fell asleep immediately.

It was just past seven o'clock when Clark shook me. I lazily opened my eyes and glanced across the room. Gene greeted me from a horizontal position while stretched out on the top bunk of the second tier. Beyond Gene's head, I observed that my bunk had been made up with blankets turned down.

While I had slept, Clark had prepared my bed and Rhinehart

had prepared the evening meal. Rhinehart went on the stove at 6:30, and was out of the kitchen promptly at 7:00 when combine C's time was up. He and the K.P.'s were entering the room carrying three vessels and a coffee pot as I got out of bed. The table was set with twelve tin plates and cups, made from tin cans. The twelve sets of G.I. silverware were furnished by the American Red Cross. Rhinehart divided the food equally on the twelve plates and passed the empty cooking vessels to combine "B." He had prepared mashed potatoes, gravy, spam, two pieces of black, toasted German bread, and coffee. Clark called the combine for chow, and we took our places about the table. Clark and Pirtle sat at the ends of the table. When the combine was in place, Clark bowed his head and said grace.

The meal was conducted in an orderly fashion. Every man used manners that would have made Emily Post proud. I was full of enthusiasm, but I must confess that I could not overcome the strange feeling I received from Peter Glass's sitting across from me. I kept staring at Glass and wondering how white men could be so close and friendly with a Negro. They acted as if there were no difference whatsoever. This was my first time to eat with a Negro, and I am afraid that I resented Glass's presence. I tried not to watch him, not wishing to offend him, but I found it difficult to control emotions derived from my environment for twenty years.

Rhinehart looked at me and Gene with envy. "You were lucky not to be required to meet appell."

"What in hell is that?" Gene asked.

"Roll call, in German. It usually lasts twenty minutes or longer. We have it twice a day, in the morning and the evening at a set time. The Germans are funny. They want to make sure we are all still here."

Clark interrupted. "The German Commandant excused all new prisoners today, so you could rest. You will meet the formation with the rest of us in the morning. Pirtle, after appell in the morning, I want you to come back here, so you can explain everything to Ken and Gene, and assign them their duties on the roster."

"Roger," Pirtle answered, as he continued eating.

After dinner Mott and Swanson cleared the table and washed the dishes. The rest of the men lay about the room smoking and talking. Finally, Clark asked Gene and me to join the combine at the table for a little talk. When the K.P.'s had taken their seats, he began:

"You men have a great deal to learn, and we will start now by telling you one of our greatest secrets." Clark was interrupted by a man at our open door.

"Clark, your combine is third tonight."

I wondered what he was talking about. The man disappeared quickly, and Clark looked at us.

"Don't look so surprised. In every combine in this compound men are assembled as we are doing now. The Germans lock us in the barracks at 9:00 p.m. About that time the Sagan news will be read in this barracks. Each block has its own readers. That was one of them telling us that we would be third to hear the news tonight. Tomorrow we will be fourth and so on, rotating our position each night. A reader will appear in the combine while we are gathered around the table. We will have cards and books about, reading and playing cards. He will join us and read the news. If at any time a 'goon' enters the barracks we will be signaled by the call, 'Tallyho.' It was taken from 'Tallyho the Fox,' and in Kriegie language it means, 'Tallyho, the goon guard.' By the way, 'Tallyho' is used at any time any German enters any barracks. Each combine continues the call, passing it down the line to the next group.

"When the reader takes a seat the cards will be dealt, and we will start playing poker. You men must remember to act naturally if we are raided. Just do as the older men do, and we won't have any trouble. While the news is being read we will have a man on duty watching in the hall. I want you to know that the news is taken from the British Broadcasting Corporation, known to all of us as B.B.C. It is official, and we have our own radios. Never ask about the radios. When and if Intelligence wants your help they will call on you. The Germans know we are getting the news. Therefore, it is imperative that you never mention the news at any time to anyone. Never discuss it with a friend, don't mention it in the latrines; don't talk about it in the barracks, outside, or anywhere you may happen to be. All German guards and ferrets speak perfect

English, and they are plenty smart. They were chosen for their jobs. They are now doing everything possible to discover our secret. Ask no questions about when it might be read. If it isn't read some night, that will mean our Intelligence has been tipped off about a raid. Now is that quite clear to both of you?"

"Yes," we both answered.

Clark continued. "I realize that you will be curious at first, and will want to discuss the news. We all want to talk about it because it gives us great hope. Sometimes even the best men will slip, in spite of hell, but a mistake here is more costly than you can realize. We pay dearly for every mistake. If a man or five men are caught by our Intelligence discussing or even mentioning the news, the entire block in which they live will be omitted from hearing the news for four days. This is done to strengthen security. So you see one man can cause a hundred men to miss the news for four days. I want to warn you that if you should slip, and if you are caught, the entire block will consider you an outcast. None of the members of your block will speak to you until the news is read again. So keep your mouth shut. By the way, the news is referred to as 'Dave.' After 'Dave' has been read we will discuss it for exactly ten minutes among ourselves. After we leave the table it will not be mentioned until the following night. The news is read any time from seven until ten."

I was about to speak when a young lieutenant wearing shorts came through the passage and took a seat at the table. Pirtle dealt the cards and cigarettes were placed on the table. The lieutenant opened two sheets of white paper, and began to read in a low voice.

"Our Intelligence warns once again that 'Dave's' security is being broken. Block 10 has been restricted from 'Dave' for four days. This must stop at once, as the goons are aware of 'Dave,' and are trying at every turn of the clock to learn 'Dave's' origin. Starting tonight if 'Dave's' security is broken by a single man, his entire block will be suspended for one week. It becomes the bounden duty of all other blocks to carry out the silent treatment with those on suspension. Let us cite you an example of what happens when men talk. Last night the German ferrets and guards raided barracks 147 at 8:15. 'Dave' was being read and 'Tallyho' was signaled throughout the barracks. The reader burned the commentary, but we came within inches of being caught. Once they know for sure that we have radios they will wreck the camp to find them and will probably call in the Gestapo. Starting tonight, by order of the Commanding Officer, if any emergency arises the reader and those about him will chew and swallow 'Dave' without hesitation.

"*Announcements:* General Vandermann has it from the grapevine that the Gestapo may visit us next week. If this visit takes place we will be tipped-off two days ahead of time. Conditions must appear terrible. The actors will be given their parts in advance. Food rations will be cut temporarily. The hospital will be filled. You will salute every member of the Gestapo while they are in our compound.

"*West Front:* The news remains the same. General Patton's army is reported to be bringing up reinforcements for an apparent drive toward the Rhine River. The First and Third Armies have advanced two kilometers in four sectors. The Port of Brest is under attack from all sides, but the Germans are still holding out.

"*Russian Front:* The Russians are still mopping up on a one hundred-eighty-mile front in the Warsaw area. Large reinforcements are being moved up along the entire Russian front. Advances have been light, carried on as scrimmages to find out the German weak spots.

"*Air War:* The American Eighth and Ninth Air Forces launched a heavy attack on the Ruhr area, bombing Dusseldorf, Koblentz, Kassel and Cologne. Over twelve hundred bombers and eight hundred fighters were used in the attack. Great damage was reported everywhere. It has been announced by the citizens of Cologne that they will hang airmen bailing out over their city.

"*Inside Germany:* The German war of nerves grows with shortages of food and material. The expected Russian attack has caused much talk and is a threat to Germany's bread basket. A civilian army is being organized to fight the Russians.

"That's it."

Most of the men left the table disappointed by the news. I wondered when the expected attack would come from the Warsaw area. It seemed as if we might be in Sagan longer than I anticipated. The Russians were still 280 miles away, and the Americans over 400 miles away.

XI

A Freshman Kriegie

THE REVEILLE BUGLE SOUNDED JUST BEFORE SEVEN, AND THE ENTIRE block started getting up for the morning appell formation.

"I'll be damned if I ever expected to stand reveille as a prisoner of war," Gene grunted.

Captain McGee paraded down the aisle in his shorts shouting: "Second call for appell. Fall out in three minutes."

"How much time do we have to get ready for appell, Rhinehart?" I asked.

"Ten minutes after the first bugle, and, boy, we really fall out on the double. You have to leave the barracks running."

When I finished lacing my shoes, Rhinehart handed me a cigarette.

Rhinehart was a fighter pilot, who could take an empty tin can from the parcels and make just about anything needed.

While we were talking, McGee announced the last call. "Fall out, last call, appell, everybody fall out on the double."

On the inside of the perimeter track, spaced at ten-pace intervals, the blocks fell in formation facing the center of the compound. When the combine reports were taken, the block commanders gave their blocks "at ease" and faced the north goal post.

The Senior Allied Officer and his staff were formed under the north goal post with the adjutant in front. Five minutes after the last bugle had blown for appell, he went forward, took his post and called for a report.

When all blocks had reported, the adjutant reported to the colonel. During this time the German guards were busily checking the barracks, counting the charge of quarters and sick men. The official German count could not begin until the Commandant had arrived to witness it. When the Commandant arrived, Colonel Spivey called the compound to attention and saluted. The Commandant ordered the count to be taken and we were again given "at ease." That was when I first saw Pop-Eye.

Pop-Eye was the top German sergeant of Sagan's Center Compound and he practically ran the camp. We called him Pop-Eye because he had only one eye, and he was as tough as nails. He was nearly six feet tall, stout, and had black hair. He would scream and bellow as the guards proceeded to each block to make the official German count. There were two German corporals assigned to our block, and they would always foul up the count. One would go down the front ranks counting noses, and the other would go down the rear rank counting butts. We named the front rank corporal Dumbkoff, and the rear rank corporal Shisenkoff.

When all other blocks had reported to Pop-Eye, Dumbkoff and Shisenkoff were still counting. "Ein, zwei, drei, vier," they counted as Pop-Eye appeared. He put Dumbkoff in a brace, and "ate his butt out good." Dumbkoff really got on the ball and concluded the count. Pop-Eye proceeded to the Commandant to report the official count.

When we returned to the combine, Ed Wunn came through the door with the drinking can. He told us he was going to the kitchen for our first ration of hot water. Inside, Dave Roberts was already hard at work sweeping the floor, dusting the woodwork, and cleaning the tables. Clark didn't wait for us to ask questions.

"Immediately after appell everybody makes his bed and helps tidy the room. Roberts and Wunn are serving as K.P.'s today. First thing after appell the K.P.'s clean up the rest of the combine and get our hot water for breakfast. They decide between themselves on the various jobs to be done. For breakfast we have two slices of German bread and a cup of powdered coffee. After the coffee cups have been washed and the combine passes my inspection, the K.P.'s are free until lunch. At lunch the Germans usually furnish us a cup of soup per man. The K.P.'s set the table, and pick up our soup and water ration at the kitchen. We usually toast our bread and have a thin sandwich spread to go with the soup, and, of course, we have our coffee. In the midafternoon the K.P.'s help the cook prepare the evening meal. Our best meal is at night. The K.P.'s are finished for the day when the supper dishes are washed and the table is clean. Now get busy and tidy up."

I made up my bed and studied the men in my combine. Life was highly organized, and it was impossible to be bored with this group.

Charles Swanson handed me a tin cup as we met at the table. "Ken, this is your cup. Label it because you will be using it from now on." The rest of the men gathered at the table for their cups.

Bill Mott spoke first. "Hey, Clark, have you told Ken and Gene about the fuehrers?"

"No, you tell them while we are waiting on the coffee."

"Well, each combine has a group of fuehrers. I am the bread fuehrer, and I issue your bread every morning after breakfast. We get six thin slices of bread per day by slicing forty slices to the loaf. I am an artist, and it's not easy to do. You can eat the bread anytime you want, but most of us eat two slices at each meal. Pirtle is the coffee fuehrer and he makes all the coffee. He keeps it under his bed and rations the strength and amount of coffee according to our supply. If we have an abundant supply he makes it stronger. Rhinehart is food fuehrer. He is in charge of all parcels except the coffee. He is also in charge of any German rations. He plans all meals and helps the new cooks prepare their food. He will also teach you two how to cook. Clark is the combine fuehrer. He is in command of the combine, and has the final say in all matters. If you have any gripes about anything, take them to Clark. You will never question the decision of a fuehrer. Each of us was elected by the popular vote of the combine to serve six months. A fuehrer can be removed by a secret ballot of seven members of the combine."

When Ed Wunn came back with the hot water all of the boys started banging their cups on the table to get Pirtle charged up. This aroused him, and he shouted: "All right, all right."

He got the coffee out from under his bed and filled the coffee pot with hot water. He put two level tablespoons in the pot and stirred the mixture. He then took his seat and served each man with pride. It looked like tinted brown water. But we all drank it and smacked our lips.

After coffee Byron Clark briefed Gene and me on the library.

"We have one of the finest small libraries in Germany—seven thousand books of all types. We have fiction, short stories, novels, history, English literature, business books, poetry, and even *Mein Kampf*. The library has almost any book you want and reading is highly recommended. It helps pass the time, and it does you a lot of good. Our block checks out books from nine until ten o'clock every morning. Pick out the book you want, and the librarian will fix you up with a card. A book cannot be kept over three days. If you become two days overdue, your card is taken up and you are suspended for four days. Both of you are going to be interviewed at nine-thirty, so you will have to miss the library today. Besides, you will be too busy to think about reading."

I turned from the table as a loud voice bellowed out, "Tallyho."

"A German guard just entered the barracks," Clark explained. He stepped into the hall and returned to the table.

"It's Rhoder spying on us. He is chief ferret of the camp," Clark said.

"What is a ferret?" I asked.

"A ferret is a German guard, or what we call a detective. He is a special guard who generally moves about the camp in overalls. He tries to listen to all conversation, and he constantly searches to uncover tunnels or other escape projects. We consider him a spy. He and his men speak good English, and they usually act dumb. There is nothing to worry about so long as you understand and stay smarter than the ferrets."

Rhoder walked through the barracks chatting with the Kriegies and talking cheerfully about everything. When he got to our combine he came in and sat down at the table.

"Hello, Clark. I was talking to a friend last night from Berlin. He told me that when we win the war the Fuehrer is going to bring all of you boys to Berlin to rebuild every stone of the city."

"Really," Clark answered. "Tell him we said we will be glad to accommodate, *if* you win the war."

Rhoder turned to me with a smile and said:

"How did you like the news last night?"

"What news?" I answered.

"Dave, the British Broadcasting news, that is read every night."

"I didn't hear any news. In fact, I didn't even know that you gave us the news." I looked at Clark as if I were entitled to an explanation.

"Ve-ry clever," Rhoder said. "You really train the new ones fast around here. This boy will be ready for your intelligence corps in a few more days."

Without another word he left. In a few minutes Clark went to McGee's room.

"So he questioned Kenneth about the news," McGee said. "Well, he doesn't know a thing. He is just trying to get a lead from one of the new men. By the way, send Coletti to the kitchen and Simmons in here. Thompson's men are on their way over and they want to interview both of them."

"Roger," Clark said.

These men were a part of Thompson's Intelligence who were assigned to maintain the camp record. They were mostly news readers from the various barracks. As I entered the room, McGee introduced me to First Lieutenant Steve Barton.

"Well, Ken, I won't waste time with formalities as this will take long enough as it is. Tell me your story from the day you were shot down until you arrived at Sagan."

I related the entire experience while Barton listened and took notes. Occasionally he would interrupt to ask me a question or to discuss some particular incident. When the interview was over, he lit two cigarettes, gave me one, and closed his notes.

"Ken, you are now a part of our camp, and you are therefore entitled to know the compound and camp secrets. Have you been instructed on the security of Dave?"

"Yes, Clark, my combine leader, gave me a long lecture."

"Good, then we can skip some of the minor details. Probably to your suprise, we have one of the best intelligence systems ever known inside a prison. The news is read to every man in the camp and every word of it is official. You will never know where it comes from, but you may be called upon sometime to help us carry out the complicated method of obtaining it. If we ever should call on you, remember to carry out your orders and ask no questions."

He studied me several moments as he paused and smoked. Then he continued:

"Besides the news we have many projects going on such as the camp record. This record is highly secret, and it is composed of documents listing mass atrocities and single atrocities committed against Americans. It also describes important military targets in Germany, complete reports on German internal conditions, and secret military orders stolen from high ranking German officers. Many men work on this record, and its security is of great importance.

"We are also engaged in digging several tunnels for future escape. All of these tunnels are planned and supervised by Intelligence. We have a complete forging department composed of skilled artists who make fake passports, false documents, compasses, maps, tools, and other needed articles.

"In addition to this, we have a secret trading post. It is composed of Kriegies who speak perfect German. They are constantly engaged in bribing goon guards. These secret traders exchange cigarettes, chocolate, and coffee, for paper, radio parts, magnets, pins, ink, saws, hammers, nails, hatchets, and maps of all kinds. All of these agencies are branches of our Intelligence. We have a total of five hundred men in the camp working daily to carry out the various projects. Now you can better understand why we must be so careful about incoming prisoners, and in maintaining security among ourselves. Once some major item has been discovered by the Germans, mass searches could uncover our entire security program."

He got up to leave and said: "You will like Sagan, Ken, if you put out as much as you get in return. Goodbye for now, and good luck."

After the K.P.'s had cleared away the noon meal dishes, Tom Pirtle and Bill Mott challenged anyone in the combine to a game of bridge. Bob Rhinehart asked me if I played and I told him I had been playing all my life.

"They are sharp players, but if you are willing, that's good enough for me," Rhinehart said.

When Mott suggested a fifth of a cigarette a point, Bob Rhinehart nearly strangled.

"Wait a minute, Mott," he said. He looked at me and said, "These guys are as good as they come, and they will break us unless you really play good bridge."

"A fifth of a cigarette per point is satisfactory with me," I answered. None of them had expected me to make this statement since I had never seen them play, and since I had been informed that they were experts. Pirtle welcomed the opportunity to strip me of my cigarettes and teach me the lesson that old Kriegies were superior in everything.

We agreed to play three rubbers and quit. Mott and Pirtle won the first two rubbers, and at the end of the second rubber Rhinehart and I were two thousand points behind.

Before the third rubber Rhinehart and I discussed our bidding and formulated a set policy. We started picking up excellent hands. The first hand I bid a little slam, Pirtle doubled and we made it. During the next few hands we had game several times, but Pirtle lost his temper and overbid trying to keep us from getting game. We doubled, and set him three times for nine hundred points. We finally made game, and won our seven hundred rubber.

When the game ended, Pirtle was enraged by the defeat. We had beaten them 3150 to 2100. This was 1050 points difference, or ten and a half packs of cigarettes each. Pirtle had suffered his first major defeat in over four months, and, worst of all, by a new Kriegie who was playing his first game of bridge at Sagan. He went to his locker, counted out the cigarettes and slammed them down on the table.

After the evening appell Clark asked me to join him in a walk around the perimeter track. We made three slow laps and talked.

We reached the barracks just as the lights came on and a whistle blew signaling official darkness.

"That's the black-out whistle," he told me. "They blow it when they turn the lights on. We close the blackout shutters, and we must walk in special roped-off paths to the toilets, library and theater. The outside toilets and library are closed at 8:40 p.m., and we are locked in for the night at 9:00 p.m. We have an inside latrine for night use at the end of the hall. Sometimes we have stage plays or musicals at night in the theater. Different blocks are scheduled at different times, and we are allowed to stay out until 11:00 p.m. on these special nights. By the way, taking daily walks is really good for you. You will learn that exercise makes you feel better. The lack of sufficient food causes poor circulation, and you need all the exercise you can get."

"How much weight have you lost, Clark?" I asked.

"About thirty pounds in all."

"That seems like a lot to lose."

"It is a lot, and that is why you also want to bathe and keep clean. Wash your clothes twice a week. You are more susceptible to disease and you need to give your body the very best care possible. Most of us sleep twelve hours out of every twenty-four. That may sound like a lot of sleep, but there are good reasons for it. You preserve your strength which helps to make up for the lack of sufficient food. You pass the endless hours of time and enjoy those hours you are awake; you are less nervous and irritable, and you are not as hungry because you are not burning as much energy. That is why I told you to follow the old Kriegies. They have tried everything, figuring every angle, and their methods are usually wise."

After dinner Pirtle insisted that the combine play a game of Monopoly. He got the game out and started spreading it on the table. Tom Pirtle's home was Nashville. He had been a prisoner for twenty-one months, and had become extremely pessimistic. He had a heart of gold, but now he refused to build up his hopes about anything.

Each member put up ten cigarettes and winner was to take all. I was the first man to hit on Boardwalk. Pirtle became excited as he ruthlessly foreclosed on my property and put me out of the game. With each throw of the dice he was as excited as any child who ever played the game. He built hotels everywhere, and pounded the table as he eliminated each man. I never knew before that Monopoly could be so exciting.

That night there were no new developments in the news. The block retired to bed in silence and disappointment.

By Friday, November 10, 1944, I had begun to fall in line with life at Sagan. I had visited the library several times, completed a full week's washing, made the shower parade, pulled K.P., and walked the perimeter track over fifteen times. I had developed a general knowledge about the compound and had learned a great deal about its many functions. In my spare time I helped Rhinehart prepare several meals, and tried to learn something about cooking. I read and slept every afternoon, and offered my services whenever called upon.

About this time I resolved to keep a diary of the strange experiences of this new life. I determined to make it accurate and reveal each incident as it actually happened. Documents of this nature were forbidden by the Germans, and punishable by severe penalties. Searches were conducted regularly and every effort was made by the German goon guards to uncover hidden information. The goons were excellent detectives, but I soon learned that their greatest failure with American prisoners was their underestimation of American ingenuity. Within a week I solved my problem when I hit upon the idea of writing my memoirs on toilet paper issued through the American Red Cross.

I started one day when the first roll of toilet paper was issued with the food parcels. I removed the outer wrapping paper and unrolled the first fifty sheets of tissue. I then rolled these sheets into a smaller roll and started writing toward the larger one, maintaining two unbroken rolls of paper. When I finished writing each day, I re-rolled the larger roll leaving its outer surface blank. I carefully replaced the outside wrapper and placed the roll on a shelf in open view among my other meager possessions. For about two hundred days I was to follow this procedure.

XII

Surprises

ONE DAY BILL MOTT OFFERED TO GIVE ME A HAIRCUT.

Bill was from New York City, and had a real sense of humor. In civilian life he had been an artist, and could paint, carve, or sculpt anything. He wanted every man in the combine to grow a mustache, and I was his eighth victim.

"If you will grow a mustache I will give you the best crewcut in the camp, trim your hair regularly, and trim the mustache."

"I am willing, but I don't think enough hair will grow under my nose to make a mustache."

"Oh, yes, it will. Look at Clark. He is blond, and has one of the neatest mustaches in camp. It took four months to get it just right, but it is truly a work of art."

He got out all of his equipment which included hand shears, scissors, comb, brush, and razor. He placed me in a straight-back chair by the window, and went to work.

When Mott finished, I looked in the mirror and saw the best haircut I had ever had.

"Now, my boy, you look ten years younger. Remember, don't shave under your nose. After it grows a week or two, I will give you a trim. A month from now you will look like Clark Gable."

Peter Glass, the only Negro member of our barracks, was sitting by the adjoining window reading *Mein Kampf*. I offered him a cigarette. He closed the book and smiled.

"What do you think of Hitler's life, Peter?"

"It is very interesting; you ought to read it. What interests me most is the fact that he told the world exactly how he would rise to power and conquer Europe and the world. His plan has been here all of this time and he did exactly what he said he would do."

"As soon as I can get to it, I want to read it," I said. "Peter, where is your home?"

"Chicago. If I ever see that place again, I am never going to leave. I have no complaints about Sagan. The fellows have really been fair to me. Being one of the few Negroes in camp makes it hard on everybody."

"I know. We all come from different sections of the country and we were raised in different environments. At Sagan it doesn't seem to make any difference. If I were you I wouldn't worry about it."

"I have nothing to worry about. All of the men treat me like one of them. It is funny, though. Of all the combines in the camp, I was assigned to this one. Do you realize that eight of our twelve men are from the South? Maybe they were testing me, or maybe they were testing you guys; or maybe it just happened. In any event, I have no complaints."

We dropped the subject of race, but talked on for a while.

Friday night found much excitement and conversation in block seven. All the men were putting on clean clothes, shining shoes, and primping in every available mirror. We had been scheduled to see the second showing of *The Man Who Came to Dinner*. I thought it would be a hammy performance with amateur actors, and I couldn't imagine a band in Sagan that could play real music. After we were dressed, Dave Roberts came over for a smoke.

"You will have the time of your life," he said. "Next time you will be sweating it out like the rest of us."

I could tell they were getting restless.

"Clark has gone to the raffle. Our block will go to the second show, but we have to draw to see if we stand or sit. If we get seats we draw for aisle, row, and seat numbers. Each combine leader draws one time for twelve men."

Clark came through the door smiling.

"Well, boys, I did it again."

Loud shouts and cheers echoed through the combine.

"At ease," Clark commanded. "I drew the center aisle, row F, and seats one through twelve."

Again everyone shouted. By now I was getting excited, and I knew there must be something to it or the rest of them wouldn't be so excited.

"It is turning cold. Everybody had better wear their overcoats. The second show starts at 8:00 P.M., and it's 7:40 now."

We left the barracks in a group, and when we got to the auditorium I got another surprise. At the front door ushers escorted us to our seats. The Sagan band was playing *Star Dust*. The Sagan band was directed by Wallace Patterson. It was composed of sixteen musicians and two vocalists. I had never heard better music in my life. Clark told me that the American Red Cross had furnished the instruments, and that five members of the band had played with top bands in the states. They had copied the Glenn Miller style, and they were out of this world. For twenty minutes they played *Green Eyes, Yours, Embraceable You, String of Pearls*, and many other old favorites.

The stage manager dimmed the lights, and Wallace Patterson took the signal. They closed with *Don't Move, Just Let Me Look At You*. This was the theme song written by a trumpet player and song writer of the band. It was a beautiful ballad about a soldier returning to the girl he loved. I knew that he would sell that song some day.

For an hour and forty minutes I felt I witnessed one of the best plays ever to appear on a stage anywhere in the world. Packed with laughter, drama, and superb acting, we were spellbound in our seats. In the final act of the play I again learned something new about Sagan. I was living among artists, actors, musicians, directors, and talent of all types.

The next morning was Saturday. After appell we returned to the barracks, ate breakfast, cleaned the barracks thoroughly, dressed in our best clothes, polished shoes, and prepared to stand the Saturday morning inspection. When call to quarters sounded, each man took his place in front of his tier and awaited the inspecting officer. Each week Colonel Spivey and six colonels on his staff divided the barracks for the inspection. When the inspecting colonel entered our barracks, Captain McGee called us to attention and saluted. Colonel Smith was the inspecting officer and the older Kriegies knew it would be a pushover. Colonel Smith proceeded to each combine inspecting each man's personal appearance, closet, bed, the kitchen and general appearance of the block. When Colonel Smith finished he gave us "at ease" and proceeded through the center hall door to inspect Block Eight, located in the other half of our barracks.

I didn't know we would have inspections in prison, but I had been at Sagan long enough to realize that it was good for morale.

The Saturday night reading of the news was dull, bringing no new developments; but following the news, eyes opened when a special message was read from Colonel Spivey.

"Gentlemen: All work on George will cease immediately. The Germans have been tipped off somehow, and the ferrets will be on the warpath. No one will mention George until further orders. Any hidden dirt will be dumped in the latrines the first thing in the morning.

"A search is expected to take place early Monday morning, so all news from Dave will be discontinued until further orders. If you get cold standing outside in the snow for hours during the search; and if you get lonesome for Dave, then remember this: someone talked or opened his mouth at the wrong time. That someone caused this search to take place, and that someone made it necessary for us to suspend the news." Signed Colonel Spivey.

Loud moans and groans followed the colonel's message, for Dave was one of the greatest expectations of prison life. A search was a very serious matter, and the older Kriegies were aware of the great discomforts of it. Gene looked at Clark dumfounded, and asked:

"Who is George?"

"George is our newest tunnel. It's being constructed from the kitchen of block one to the woods west of us. 'He' is the finest of his kind, being twenty-two feet deep, five feet high and three feet wide. 'He' will be a permanent structure, and when finished 'he' will extend to the woods four hundred feet west of barracks one. We intend to bring 'him' to the surface forty-eight feet inside the wooded area. 'He' is only twenty-seven feet long, but we are working slowly and carefully so as not to arouse the suspicions of the ferrets. When George is finished, 'he' will not be brought to the surface until the final night for escape. That should be during the final stages of the war when the camp is under attack. If the ferrets fail to discover 'him' we will resume work when their suspicion dies out. Remember at all costs, never mention 'him' to anyone."

Sunday morning Catholic Mass was held in the auditorium from six until seven. Short appell followed at eight, and breakfast was over by eight-thirty. The Protestant services were held from nine till ten, and from ten until eleven. Two Protestant church services had to be scheduled because of the large crowds. The first service was held for all Protestants in blocks one through ten, and the second service was for those in blocks eleven through twenty. The schedule rotated every other Sunday. All men enjoyed the first service most, because the general attended it, and he usually led the morning prayer.

Church attendance at Sagan was indeed different from attendance in the Army or in civilian life in the United States. Ninety-eight per cent of the men attended one of the three services. It was a major part of every man's life, and men looked forward to church as the number one event of the week. Even with three different services several hundred men had to stand, in order to attend the service, but it was unheard of for a man to complain because he didn't have a seat. Many Protestants who were scheduled for Sunday morning details attended Catholic Mass at the early service. Though there were fewer than three hundred Catholics in the center compound, early Mass usually brought over seven hundred men through the church doors for worship.

Sunday was a day to worship God, and it did not end with church services. Men studied the Bible individually and in groups in the combines. This studious atmosphere about religion was something I had never met before, and will probably never meet again.

At 8:45 A.M. on Sunday, November 12th, I attended my first church service at Sagan with Byron Clark, Bill Mott, Tom Pirtle, Ed Wunn, and Dave Roberts. We arrived just before nine, and found six seats near the second aisle. By nine the church was packed, and the standing space was nearly filled. I observed that eight center seats on the center front aisle were tagged with white slips marked "Reserved."

Bill whispered: "These seats are for the general, Colonel Spivey and the other six colonels. They always attend the first service. We don't stand for the general, since we are in church."

I kept looking at the beautifully decorated stage. The pulpit was draped in a white silk cloth with gold embroidered outline. The fifty choir chairs were covered in black cloth, and the choir section was roped off by two small ropes painted white. The red velvet stage curtains had been pulled back and draped to cover the side walls. A large blue cloth swung wavelike from the ceiling in front of the stage to the back wall. The center of the back wall was covered with the American flag. The flood lights from the stage swept over the empty choir section, and beamed their brilliant rays on the red, white, and blue background of Old Glory.

When the general and his staff entered the church, it was filled. The ushers closed the three exit doors, and a hush came over the auditorium. This was my first view of General Vandermann, and I shall always remember it. He was in full dress uniform, wearing the Air Force uniform with forest green battle jacket. His four sets of silver stars glittered. The general had rugged features and a stocky build. His face was dark and full of character. In his early fifties, he had a high forehead, coal black hair, and a well-trimmed black mustache. He walked to his seat like the head of West Point.

Chaplain Daniel now mounted the stage and seated himself in the straight-back chair behind the altar. The choir section was still empty, and I whispered to Tom Pirtle: "Where is the choir?"

"Behind the stage. Just wait, and you will see."

Chaplain Daniel, dressed in pinks and officer's blouse, rose and stood in the pulpit.

"Let us pray."

We rose from our seats with heads bowed as he began to pray. As he prayed, a beautiful choir of hidden voices began to sing softly *In the Garden*. When the prayer was over, the hidden choir started faintly singing "*O Come All Ye Faithful.* The faint music grew in volume as the side curtains were drawn, and fifty men garbed in white choral robes entered from both sides of the stage. The volume of voices steadily increased, blending in perfect unison. When the song ended the choir took their seats.

General Vandermann rose from his seat and turned sideways to the audience with his Bible in his hand. In a warm rich voice that filled every part of the church, he read the Beatitudes. After "Blessed are the peacemakers; for they shall be called the children of God," he paused and looked over the audience. Raising his voice, he continued. "Blessed are they which are persecuted for righteousness' sake; for theirs is the kingdom of heaven. Blessed are ye, when men shall revile you, and persecute you, and shall say all manner of evil against you falsely, for my sake. Rejoice," he shouted, "and be exceeding glad; for great is your reward in heaven, for so persecuted they the prophets which were before you."

Slowly he closed the Bible while looking into our eyes, and took his seat. There was a total silence in the auditorium.

Two songs followed, *The Battle Hymn of the Republic* and *Jesus Is Calling*. The entire congregation sang.

Again Chaplain Daniel appeared in the pulpit and bowed his head. "Let us pray." As he led the prayer, hundreds joined him: "Our Father which art in heaven. . . ."

Solemn, stern voices echoed throughout the church and over the silent grounds of the compound. Following the prayer, without introduction, the choir leader rose facing the audience and announced: "We give you our special arrangement of *The Holy City*." Once again the beautiful blend of trained voices vibrated within the church walls.

The sermon was twenty minutes long. When the sermon was concluded we closed with *Just As I Am* and departed through the east and south doors. Outside there were five hundred men standing at the west door waiting for admittance.

XIII

The Goons Search the Camp

DURING THE EARLY HOURS OF MONDAY, NOVEMBER 13, A HEAVY snow began to fall, forming a solid white coat over the grounds of Sagan. By dawn the white coat of snow lay several inches deep and the temperature continued to drop.

Chief ferret Rhoder entered the Center Compound early with a staff of twenty specialists and the entire German guard of the Center Compound. He spoke to Pop-Eye for several minutes, outlining his plan in detail. Pop-Eye barked out orders placing his guards at every entrance to every barracks in the compound. Pop-Eye personally checked every guard under his command, and ordered the bugler to sound appell.

Promptly at eight o'clock, Rhoder ordered every barracks to stand by for a search. He was extremely cocky as he marched down the aisle of our barracks while we dressed. He loved the element of surprise, and he thought he had caught us off guard. But our Intelligence had paid one of Rhoder's men fifty packs of cigarettes for the tip-off about this search, and we were ready.

We were lined up in a column of two's as soon as we finished dressing. Four guards stood at each door and each of us was thoroughly searched from head to foot as we passed through to the parade ground. A fifth guard stood outside, and counted us as we came out. When all the barracks were empty the ferrets totaled the count, at each barracks door, for the appell. Pop-Eye moved all prisoners to the parade ground, and placed guards on the outer edges of the perimeter track to keep us there.

I noticed that four guards were placed outside at the four

corners of each barracks. Rhoder divided his twenty specialists into five teams, and the search began. Closets were emptied, beds were removed, and pally-asses were gutted. Kitchens were taken apart piece by piece, and food parcels were checked item by item. Floor boards were removed and examined for possible tunnel entrances. Several sections of walls were ripped out in every barracks in search of possible hidden weapons.

When the inside search was completed, the search teams moved outside. They checked the barracks roof, attic, foundations, and outside walls. Several sections of outside walls were torn loose and checked. All of the loot was loaded on two wagons Rhoder had brought into the compound.

From the barracks the search continued to the latrines, kitchens, library, and auditorium. Rhoder had several men dip buckets attached to ropes into the latrine holes to check for hidden articles. He went over the Center Compound with a fine tooth comb examining everything he could think of. He had made a list of places he had apparently failed to check before, and he checked off the list as his men completed the search. He hunted madly for tunnels after all buildings had been searched. He made three special trips back to the latrines, library, and auditorium. He must have stayed in the auditorium for an hour. The search lasted for four and one-half hours.

During all of this time we stood in dozens of small groups talking, smoking, and moving around trying to keep warm. The snow continued to fall, and all of us grew anxious, nervous, and tense as the search failed to end. Those responsible for vitally important projects and secret material suffered agonies as the minutes slowly passed by. Hundreds of hopeful hearts anxiously awaited the end of the search, to learn if a month's, six months', or even a year's work had been in vain.

The Germans never failed to uncover a few hidden weapons, tools, and documents, but this was because we always placed them where the Germans would be sure to find them. The German ferrets had to find something for the Commandant, and the quicker they found a few saws, shovels, knives, and misleading secret documents, the quicker the search would end. We saved valuable equipment by making sure they would find enough junk to satisfy their Nazi ego. We did everything possible during a search to make the Germans believe they were superior in wit, and it paid off in big dividends.

Preparing valuable equipment for a search was of great importance. It was the responsibility of Intelligence to hide this equipment so the ferrets would not find it. Some of it was placed in sight of the search teams. Radio tubes were placed in the center of a roll of toilet paper, and the toilet paper was left on top of the clothes locker. Some small tools and weapons were placed inside of G.I. shoes in the closets. Other items were left in open view of the ferrets. The most valuable items had to be hidden where they could not be found. Closets were rebuilt with false bottoms. The Nazis never thought of tearing a closet to pieces, since their carpenters had built them. Stoves were reworked with secret compartments, and the concrete foundations for the barracks stoves were hollowed out by expert masons. Many items were buried on the parade ground the Sunday night before the search. They never thought of digging up the parade ground. It was surprising how many items they would miss that were in plain sight. Major Thompson once said that the only radio equipment ever destroyed was a receiver set hidden in a pasteboard box, which was crushed by a German ferret when he sat down to rest.

Finally Rhoder came out from under the auditorium, and blew his whistle. The search was over, and all discovered items were taken by the search teams to the two wagons. Clark, Pirtle, and I joined several hundred prisoners who gathered about the wagons to see how much the Germans had found. Rhoder always allowed us to get a good look, to show us that he was the smartest man in Sagan. When we gathered about the two wagons, I got another surprise. Special actors stepped forward to express their utter amazement while others flattered Rhoder and his assistants about how he had outsmarted us. While they kidded Rhoder and worked him into a stage of hysteria, sleight of hand artists stole dozens of articles from the wagons, and passed them through the crowded ranks.

During this repossessing process, I saw one of the boldest acts ever tried at Sagan. David Bowling, one of the actors, waited until the ferrets started bragging and boasting of their cleverness. He suddenly rushed forward, slapped Rhoder's assistant on the back, grabbed a large folder of maps, and walked through the crowd to the barracks. Bowling passed ten German guards with the bundle under his arm, yet not one of them ever dreamed that he was carrying loot from their wagons. I decided that the bolder the act, the greater the chance for success.

When Rhoder finally assembled his twenty detectives and prepared them for departure, nearly one-third of our possessions had been regained without his slightest suspicion.

Late in the afternoon, Captain McGee announced that projects Dave and George were both safe, and that we had lost only what we had intended to lose. He also informed us that forty-seven items had been repossessed from the wagons.

All of us were exhausted and half frozen from the long search, but very happy over the results. I had never been around so many clever men in my life. I decided that Kriegie Intelligence and Kriegie strategy could not be equalled anywhere in the world.

During the next three nights, Rhoder launched full scale surprise raids on nearly every barracks in the compound, in an attempt to uncover something that would lead to Dave. Since the news had been suspended, every attempt failed. Sad faces and grouchy dispositions convinced Rhoder that his Kriegies had lost their secret news agency. He finally gave up his overtime night activities.

On Friday, November 17, at the usual time, the readers of Dave resumed their old duties, and the news began once again in the compound. Mott handed three men a book to look at, and Pirtle dealt six hands of poker, as the reader joined our table. He concluded with a message from Colonel Spivey:

"Congratulations on your excellent cooperation and efficiency during the search. You displayed fine teamwork after the search, and recovered much property. Most important of all, not one single major project was discovered."

We sat at our combine table for the full allotted ten minutes discussing the news of the war's progress. All members of the combine were the happiest I had seen them since I came to Sagan. Adding to the success of the search just completed, many old prisoners said it was the best week they had experienced since they had been captured.

During the second and third weeks of November, full scale winter reached Sagan, and packed the snow a foot and a half deep. We were confined more and more to the barracks, and daytime winter activities continued to grow. Most of the members of my combine spent their time reading. There were many Kriegies who would read a new book every day.

There were some books that could not be checked out of the library, and I spent as much time in the library as I did in the combine. I devoted many hours each day to reading books of all types. The more I read the greater the desire I had to read. I spent several days studying the works of Shakespeare, Byron, Keats, Shelley, and Samuel Johnson. I became so interested in Johnson's works that I read Boswell's *Life of Johnson* from cover to cover. I also read biographies of Henry Clay, Lincoln, Napoleon and Henry the VIII, and I read *Mein Kampf*. I read, too, several popular novels and, believe it or not, a history called *Economic Life in Europe*.

I, like many another new Kriegie, had found the starting point on a long road to a new and different life. It was strange, bold, daring, and almost unbelievable; but the remolding process of Kriegie Kenneth Simmons had started. I was exchanging old habits, ways of living, principles, and ideals for new ones that would remain a part of me the rest of my life. I was, of course, unaware of a change in my character, but it was slowly taking place.

A week before Thanksgiving the combine leaders of the entire compound were called to a meeting in the auditorium with Colonel Spivey. Since the first of September, one-eighth of the total rations of every combine had been saved for the Thanksgiving and Christmas holidays. We called this saving of food "saving for the bash." The "bash" was the day of celebration when we would eat until we could hold no more. The meeting was called to decide whether we would "bash" part of our reserves for Thanksgiving and part for Christmas, or whether we would have a mild Thanksgiving and have a tremendous "bash" during the Christmas holidays. A large majority of the combine leaders voted for a mild Thanksgiving and an all-out "bash" and celebration during the Christmas holidays. We had nearly two extra weeks of parcels and food supplies already saved, and everyone agreed that this would be the most elaborate Christmas ever known at Sagan.

Thanksgiving was celebrated with special church services, and a very good noon meal.

On November 30, the "Annual Center Compound Bridge Tournament" started, and every combine entered two teams of players. There were 160 combines in the Center Compound, and 320 teams filed. The filing fee was one pack of cigarettes. One hundred judges were appointed to supervise the 320 competing teams.

The battle of wits began with several hundred teams scattered over the compound in bitter four-man contests. Combine C entered Bill Mott and Tom Pirtle as Team 7-C-1 and Bill Rhinehart and me as team 7-C-2.

The German guards were carried away by the tournament, and even Rhoder tried to watch at least one game every day. There were no games scheduled in block one, and project George was resumed under the most ideal conditions. Two hundred men started to work on George as the tournament progressed.

By the end of four days' playing, 180 teams were eliminated, and over half of the 140 remaining teams had lost one game. The individual battles continued in every block of the compound, and the judges kept score, settled debates and eliminated the losing teams. By December 4 only 50 teams remained, and the semi-finals started.

Combine C was the only combine in the compound with both of its teams still in the tournament. Rhinehart and I had lost one game while Mott and Pirtle were undefeated. The men of my combine boasted all day Sunday that we had more brains in Combine C than any barracks in the compound.

During the two days of semi-finals Rhinehart and I were eliminated, but Mott and Pirtle reached the finals with only one loss. When the finals started, large crowds gathered to watch each game. The judges warned that all observers must remain silent, and that any friendly suggestions or signals to a team would cost that team a 500-point fine. With six teams left in the tournament, Mott and Pirtle were eliminated in a close three rubber game.

Thursday night, December 7, 1944, the championship game was played in the library at 7:00 P.M. Rhinehart, Pirtle, Mott, and I decided to go early in order to get good seats. Dickins and Jones of block eighteen were playing Carleton and Caver of block eleven.

When we arrived at the library, a special table had been set up, and four judges were chosen to supervise the game. The library was jammed with spectators, and the judges demanded absolute silence. In vain, for every play of the game brought noisy comments from various bridge experts. At the end of each rubber, men would cheer and holler. The final rubber decided the tournament championship, and it was won by Carleton and Caver.

XIV

Operation George

FRIDAY, DECEMBER 8, IMMEDIATELY AFTER BREAKFAST, EIGHT members of Combine C were assigned to Operation George for five days. The K.P.'s, Rhinehart and Clark, were excused. Promptly at 9:30 A.M. Ed Wunn, Dave Roberts, Matt Johnson, Charles Swanson, Ted Carson, Gene Coletti, Bill Mott and I crowded into McGee's room. We took our seats on two beds, while McGee sat at his desk.

"Project George is now eighty-nine feet long. You men have been selected to work on the most important project in this compound. You must be on your toes every second you are on duty, and carry out your orders to the letter. You must at all times remain cool and calm. Colonel King is in command of George, and he will be here in a few minutes to assign you to your specific duties. You are not to discuss your duties with any other member of this block until your five days' work is over."

Colonel King and his aide entered the room. "Have you got guards outside, McGee?" he questioned.

"Yes, sir. They are building snowmen, and if a goon approaches the barracks we will be signalled."

"Good. Now, men, down to business. I am doubling the digging and removal crews to make the utmost progress between now and Christmas. The Germans are not suspicious with the Christmas holidays coming on, and this deep snow offers the most ideal conditions to dispose of dirt. Our engineers estimate that George will be four hundred feet long when we bring 'him' to the surface in the center of the woods. I am very anxious to get 'him' at least halfway finished before Christmas Eve.

"The mouth of the tunnel is located in the kitchen of block number one. Our engineers and masons have built a false bottom to the large stove in the kitchen of block one. The concrete foundation which runs from the floor of the kitchen to about two feet underground has been hollowed out. The stove rests on the four solid corners of the foundation in the kitchen. This is the mouth of the tunnel.

"We do not work at night, for Rhoder could pull a raid and take us by surprise. When men are working in the tunnel, the stove is rolled free of the foundation. Our digging crews enter the tunnel and climb down a stepladder to the bottom. The tunnel is twenty-one feet deep. It is five feet high and three feet wide. When men are working in the tunnel, the stove covers only half of the opening. The remainder of the opening is used to remove dirt. We also have an air hose in the tunnel, and air is pumped from the kitchen. We have an extension cord and light plugged from a socket in the kitchen in addition to the oil lamps in the tunnel.

"Now we have several teams of digging crews and engineers who mark off the digging. As the tunnel is lengthened by the diggers, men directly behind them load the dirt into wooden boxes. When the boxes are full, men behind the loaders push the boxes across the floor of the tunnel to the next pushers. This process is continued until the boxes reach the mouth of the tunnel. Men from the kitchen lower gallon drinking cans and buckets while men at the bottom of the mouth of the tunnel fill the drinking cans and buckets with dirt. Ropes are attached to each drinking can and bucket, and when the signal is given they are raised to the surface. In the kitchen, all dirt is dumped into pasteboard boxes or removed immediately by the dirt removing teams.

"We have many men guarding the tunnel from the kitchen, and from many areas outside of block one. We have seventy-five men assigned to removing dirt from the mouth of the tunnel. On two occasions we have called for one of the board slats attached to your bed. These boards were used to prevent cave-ins and to strengthen and support the tunnel.

"I have explained many details of the tunnel so you will understand what we are doing and what a tremendous task we are undertaking daily. All of you men will be assigned to dirt removal and guard duty. The fundamental success of the entire operation depends upon quick and cautious removal of dirt, and top-notch guards who will signal us immediately upon the approach of a German. We must always have sufficient time to drop everything in the mouth of the tunnel, close and conceal it before a German enters the barracks. You will now go to the library two at a time and ask for Captain Randolph. He will assign you to your duties."

The Colonel left immediately.

Charles Swanson and Matt Johnson were the first two to report to the library. Ed Wunn and I reported to Captain Randolph in the east reading room as Swanson and Johnson were leaving.

Randolph introduced himself, and checked our names off the list. He studied his roster for a minute.

"You two men will serve as guards today and tomorrow. After that you will be assigned to the dirt disposing teams. Simmons, block one is located in the north end of barracks 149. The kitchen of block one is the location of George. Your station will be midway between barracks 149 and barracks 150. As you know, these two barracks run parallel.

"Another guard will meet you there, and you will spend the entire day building a snow man. If you get too far along with the snow man, tear it down and start over. If a goon guard appears in the southern area you are guarding you will immediately throw a snowball at Wunn. Wunn, you will never take your eyes from Simmons. You will be stationed at the north corner kitchen window outside block one. A guard stationed inside the kitchen will be watching you from the window every second. When and if Simmons ever throws a snowball, you will sit down immediately. Whenever you sit down in the snow, that will be the signal to stop all operations. All equipment will be dropped in the mouth of the tunnel, and the stove will be rolled over the concrete opening.

"Just one more thing. Don't worry about us getting the tunnel closed if a guard starts for either end of the barracks. We have plenty of men on duty, and several methods of stopping the guards. If one doesn't work, another will. O.K. Let's go to work."

Ed Wunn hailed from Louisiana. He didn't talk much. Ed had worked with Intelligence before, and was plenty sharp.

When I reached my post between barracks 149 and barracks 150, who should be building a snow man but Ted Carson.

"They told me another guard would meet me at this spot, and we would build snow men all day," Carson said.

"Well, old buddy, I am that other man," I said.

"Boy, this is the most complicated operation I have ever seen," he said.

We started to work on our snow man facing the south. One of us kept a constant eye on the southern area while the other worked on the snow man. Ted was a very well educated boy and we had a wonderful conversation about the main goals in life before we were relieved for dinner. He was a good athlete and worked constantly to keep in good shape. Ted was very good-natured and never lost his temper, even though he had been at Sagan fourteen months.

We had thirty minutes for lunch, and both of us ate by the stove. I was just thawing out good when it was time to resume our guard duties.

At 3:00 P.M. that afternoon it happened for the first time. Dumbkoff and Pop-Eye rounded the corner of barracks 150. I nearly collapsed. I grabbed a snowball and threw it at Wunn, hitting him in the face. He fell flat on the ground, and threw one back at me. Another guard threw one and hit Pop-Eye on the head. Pop-Eye put the Kriegie in a brace, and cussed him out for two full minutes.

Five men came charging out of the barracks at a signal from the guard at the front door, and headed for the latrines. These men had cloth sacks built inside their trouser legs. The sacks were loaded with dirt and had two drawstrings attached to the sides and tied to each man's belt. The jacket covered the strings, and each man had to be careful not to overload or his legs would look like balloons. Inside the latrines, each man stood over one of the stools and released the strings. The bottom of the sack came open, and the dirt poured into the latrine holes.

Inside the kitchen of block one a major operation was completed in fifty seconds. Air hoses, buckets and ropes were dropped through the mouth of the tunnel. The stove was rolled in place, and coal dust was sprinkled all over the concrete slab. All dirt was rushed from the kitchen, all boxes were wiped clean, a card table was set up, three hands of poker were dealt, and all other men scrambled from the kitchen. The men on guard just inside the door and outside of the door did not have to perform their acts.

Pop-Eye stopped at the corner of the barracks and asked Wunn how he liked Sagan. They chatted for several minutes. I was relieved to see Dumbkoff following Pop-Eye toward the auditorium. Eighteen men were still enclosed in the tunnel working. If Pop-Eye had started toward the barracks door, "Tallyho" would have been called by a dozen men, and all operations in the tunnel would have ceased until the all-clear signal.

Several minutes after the departure of Pop-Eye, the mouth of the tunnel was opened and Operation George was resumed. Thirty minutes before evening appell construction was shut down for the day and everything was cleaned and returned to normal. Soap and water were stationed in several combines of block one. As the digging crews came out of the tunnel, several men helped each worker wash, sponge bathe, and change clothes. Each worker in the tunnel was inspected thoroughly for any signs of dirt, and laundry crews dried, brushed and cleaned the work clothes.

Saturday we had to stand the morning inspection to prevent the goons from becoming suspicious. It was nearly eleven when work was resumed on George. Ted and I built snowmen, tore them down and built them again. He talked about the great state of Iowa, and I talked about the greatest of them all . . . Texas.

"Ted, we don't do anything but tell the truth. You think we are trying to be funny, yet you have never been in Texas. We do have just about everything, and we actually have more of it. If you would ever come to Texas, you could see for yourself."

It looked as if we were going through the day without incident. It was just nearing four when I heard a commotion. I looked at Ed Wunn and saw several snowballs falling about him. He was already on the ground. Bear-Face and Shisenkoff were coming across the parade ground headed straight for the entrance of block one. There were three Kriegie guards playing with a basketball ten feet from the door. Another Kriegie, watching them pass the ball back and forth, approached Bear-Face and Shisenkoff about twenty feet from the door. Bear-Face was Dumbkoff's relief man and he was the prime idiot of the German guard.

The Kriegie guard had a cigarette in his mouth, and stopped squarely in front of their path.

"Have you got a match?" he said in high school German.

Shisenkoff gave him a light, and the Kriegie offered both goons a cigarette. They never turned one down. He stalled them as long as possible, and then told Shisenkoff in halting German: "Have you heard the news? Hitler shaved his mustache and ordered every member of the armed forces to shave his immediately."

Shisenkoff involuntarily felt of his mustache, and then shoved the Kriegie aside.

The boys with the basketball tossed it to Bear-Face. His rifle was slung over his shoulder, and he caught the ball smiling. He passed it back and caught it again. While Shisenkoff was looking about the grounds, Bear-Face laid his rifle on the ground and got in the circle passing the ball. In a few minutes, when Shisenkoff returned, he looked at Bear-Face, then the Kriegies, then Bear-Face's rifle lying on the ground. He walked up to Bear-Face and stepped on his right toe. He never said a word, but stared at Bear-Face's rifle. It suddenly dawned on Bear-Face that he had made an ass out of himself. He grabbed his rifle, slung it over his shoulder, and followed Shisenkoff to the door of block one. Everything had been cleared with minutes to spare, but ten different men shouted as loud as they could, "Tallyho."

Shisenkoff came in the kitchen and watched the men playing poker. "Always gambling," he told Bear-Face as they proceeded down the hall.

That night I sat near the stove for two hours. All of the men who had worked on George maneuvered for positions. Swanson had his rump in my face nearly the entire time.

Swanson told us of the guard who followed him to the center of the parade ground. He said he must have been walking too fast, and the guard got suspicious. He rushed into a circle of men playing ball, dumped his drinking can of dirt, and everybody started kicking snow in every direction. Swanson stayed in the circle cleaning the can with snow while a dozen men bombarded the guard with snowballs. When the goon finally got to Swanson the can was clean as a whistle.

Sunday, after church services, Combine C spent a quiet day resting for three additional days of hard work on Operation George.

Monday morning Matt Johnson and I were instructed to report to the library together. Captain Randolph again gave us our orders. "Johnson, you and Simmons will work as a dirt removing team for the next three days. You will both report to the south end of block 149. Johnson will be issued a drinking can, and Simmons, you will be issued a pair of bagged trousers.

"Johnson, you will proceed to the north end of the barracks to the kitchen of block one. They will fill your can two-thirds full of dirt. You will leave by the north end and proceed across the parade ground to the third group of men playing volley ball. When they see you coming they will form a circle. When you are covered, you will dump the dirt and wash the can with snow. The men in the circle will scatter the dirt and cover it with snow. You will then proceed to the kitchen. You will get cold water for inside washing and re-enter the south end of barracks 149. If anyone needs the water give it to them, but when no water is needed dump it in the inside latrine. Then go back for a second load of dirt. Take all of your dirt to the same group of men. On the second trip enter by the north end of the barracks. Get your third load and depart by the south end. You will make four such trips.

"Simmons, you will change to the bagged pants in Combine A of block one. Your pants will be hung in a closet. You will be properly loaded in the kitchen of block one. You will first depart at the north end of the barracks and enter the north end of the latrine. Guards will be on duty to cover you in the latrine and you will be instructed on how to operate the drawstrings. You will dump your dirt in the latrine hole and re-enter the north end of the barracks. When you are loaded you will depart through the south end of the barracks and unload in the south end of the latrine. You will also make four trips.

"Then you will change. Simmons will use the drinking can, and Johnson will use the bagged pants. Before the day is out they will stop you from using the bagged pants. We have a set ration of dirt to go in each latrine hole per day, and it must not be exceeded. After the quota is reached, you will be issued another gallon drinking can, and both of you will operate from different ends of the barracks.

"If you see a guard standing in a fixed position you will complete the trip you are engaged in, but you will not make another trip until he moves on. If the signal is given to close the tunnel or 'Tallyho' is signaled, all the men in line waiting for dirt will move swiftly to their assigned combines. Both of you will report to Combine A. Okay, get to work."

We reported to the south block commander and drew our gallon can and bagged pants. When we got to the north kitchen, there were twenty men standing in line waiting to get their dirt ration. The line moved swiftly and with ease. Three men in the kitchen removed the dirt from the tunnel, and three other men rationed it to us. All dirt going to the latrines was tabulated by loads.

During lunch, Matt and I discussed the operation, and we couldn't get over the detailed planning and the tremendous number of men working on the project. It was almost unbelievable.

That afternoon we made seven trips each to the parade grounds. It was also our first day to operate without one shutdown. The digging crews had to be changed twice, but the goons never bothered us. The goons never paid much attention to the drinking can operation, since it was nothing unusual to see a hundred men at a time carrying drinking cans of water to every part of the compound for cooking, drinking, washing, and other domestic purposes.

Of all the methods used to dispose of dirt, I thought the latrine operation devised by Intelligence was the most clever brainchild ever put into operation in Sagan. This method not only removed a large amount of dirt from the tunnel, but also removed it from the Center Compound. The five outside latrines in the compound were each used by four hundred men several times a day. Since there were only twenty seats in each latrine, congested conditions prevailed constantly. Each latrine hole was approximately twelve feet deep.

The German Disposal Department had a group of sanitary

wagons used exclusively for emptying the latrine holes, and they were constantly engaged in the process. The wagons were pulled by two horses, and a large tank was mounted on the frame of each wagon. Running from the tank was an eight-inch suction hose sixty feet long. A small Diesel engine was attached to the front of the tank. The driver would back up to the latrine door, start his engine, and run his hose to the bottom of each pit. When the pit had been sucked dry he would move the hose to the next pit. It usually required four tank loads to empty each latrine, and the sanitary wagons operated six days a week.

Since the latrines were in constant use, and moist dirt would mix well, Intelligence ordered five "gallons" of dirt to be emptied into each hole daily. There were a hundred holes in the five latrines, and five hundred "gallons" of George's dirt was disposed of each day. Like all other ingenious plans of Kriegie Intelligence, this one proved a brilliant success. Little did the Germans know that their sanitary wagons were working overtime removing three thousand "gallons" of tunnel dirt per week and serving as the greatest disposal agency that American Intelligence had in operation.

On Wednesday we served our fifth and final day's work on George. Matt and I were still on dirt removal duty. I had carried enough dirt out of George to fill a gravel truck.

The word was spread right after lunch that Rhoder was in the compound. Everybody proceeded with extreme caution and all guards were extremely alert. I finally found out what the two guards just inside the door by the kitchen were supposed to do.

Just after 2:00 p.m. Rhoder came walking briskly from the theater toward the north end of barracks 149. The snowballs flew and the kitchen guard ordered the tunnel closed. We had a large amount of dirt out of the tunnel, and a dozen men were racing to get it out of the kitchen. Several men tried to stop Rhoder but everything failed including the basketball. He marched briskly toward the north door. Just as he reached the third step, the two guards started a real fist fight. I didn't know until afterwards that it was staged. They hit each other several severe blows and came tumbling down the steps on top of Rhoder. Rhoder fell in the snow, and when he saw blood coming from the nose of one of the Kriegies, he rushed in between them and broke up the fight.

"He called me a Hitler-lover and no son-of-a-bitch is going to call me that," one of them told Rhoder.

Rhoder lectured them, and called the block commander. It delayed Rhoder for three minutes and saved the tunnel. When Rhoder finally got inside, he never even looked in the kitchen.

Wednesday evening I was delighted to get back to the combine. The eight of us spent the rest of the night sitting in groups and telling each other the thrills of working on George.

That night, when Dave was read, we learned that George was 193 feet long.

XV

Rumors of Christmas

WHEN WE RETURNED FROM APPELL THURSDAY MORNING, SEVERAL of us who had worked on George were weak and lightheaded. Working for five days in snow and freezing temperatures had sapped a tremendous amount of energy. We were eating about one-third of the amount of food we were normally accustomed to, and any exertion or exercise would tell on a man almost immediately.

When we gathered around the table for our morning coffee, Rhinehart and Pirtle were humming and joking with each other. I knew something was brewing, but I couldn't figure it out. Peter Glass arrived with the hot water, and all of us started banging our cups on the table.

"All right, all right," Pirtle shouted and grunted as he came to the table with his coffee jar and spoon. All eyes were fixed on the coffee jar. We always counted as Pirtle would level off two tablespoons and drop them in the coffee pot. "One, two." It was always "One, two," because that was always the ration. He paused and looked at us, then put the spoon back in the jar.

"Three, four, five," he shouted and dumped three more heaping tablespoons into the pot. Gene nearly fainted, and picked up the pot to smell the aroma.

"Real coffee," Gene hollered. Everybody in the combine cheered Pirtle.

"Don't get any funny ideas," Pirtle said. "McGee told me this morning that you guys worked like hell on George and I just want you to know it pays to work around here."

About that time Mott arrived with our two slices of toast, and Rhinehart came forward with a big bowl of sandwich spread and eight blocks of precious chocolate.

"I'm like Pirtle. Don't think for a minute I am getting soft-hearted. I know the eight of you are weak as hell, and need this chocolate. I'm not feeling sorry for you. I just don't want anyone trying to goof off from K.P. or cooking. You have duties to perform in this combine, and I just want to make sure you perform them. Besides, it's only a small part of our Christmas savings, and we will never miss it."

Our coffee was as precious as gold. The priceless possession in Sagan was chocolate. Several of us were so choked up we could hardly eat, to think that Pirtle, Glass, Rhinehart and Clark had given their chocolate to us.

After breakfast, Ben Mott gave three haircuts and trimmed eight mustaches, including his own.

"Well, Ken, you said it wouldn't grow. Look at it now, boy. Outside of that small gap on your left upper lip, it's a masterpiece."

Byron Clark, Tom Pirtle and Dave Roberts were gathered around the stove telling tales about Pop-Eye when Charles Swanson came storming through the door.

Swanson was the youngest member of the combine, only nineteen. He was also the most gullible. He never looked where he was going, and usually crashed into two or more objects every time he moved through the barracks.

He hurried through the door and stepped on Roberts' toes as he elbowed his way to the stove and stood panting, trying to catch his breath. He seemed extremely excited. He carelessly removed his coat, swishing it in Pirtle's face and knocking loose snow down Tom's neck.

Clark laughed as he looked at Swanson's red cheeks. "Let's hang a bell from the ceiling," he said. "The first man to see Swanson coming can ring it and give us time to clear the room."

"I'm sorry, fellers, but I've got the hottest rumor in camp. I just heard that eight box-car loads of Christmas parcels just arrived from the Red Cross, and that the General is going to give each man a full Christmas parcel plus our regular half parcel for both Christmas and New Year's weeks."

Mott had just finished trimming my hair. "What was that?" Mott asked excitedly.

"What latrine stool did that one come off of?" Pirtle asked.

"That's the kind of rumor that hurts morale," Clark joined in.

"It's not a rumor. I heard it twice in the last ten minutes. I was in the latrine and . . ."

Pirtle interrupted. "I knew it, I knew it. They always start in the latrine."

"Hell, let him finish," Mott shouted. "At least it's a good rumor, and that is more than most of you bring in."

"Well, I heard two of the mess sergeants tell Captain Ferrell, who was sitting three seats down from me, that eight boxcar loads of Christmas parcels had arrived, and the goons wanted them unloaded before dark. They told Ferrell the colonel had sent them to find him; and since he is in charge of all parcels, I have every reason to believe it."

Ted Carson put his arm around Swanson, and said: "I believe every word of it. Don't ever fail to tell us a wonderful rumor like that."

"I thought you said you heard we would get a parcel and a half each," Pirtle prodded.

"Let me finish, will you! One of the sergeants told Ferrell that the box-cars were loaded with sixteen thousand five hundred parcels. He said there were over two thousand parcels to the car. He also told Ferrell that there were enough parcels to give each man a whole Christmas parcel besides our regular half parcel. Ferrell got out of the latrine before I could leave. When I came outside by the bulletin board, everybody was talking about it. The news really got around fast. Ferrell must have told them because several told me the general had decided to give us a whole Christmas parcel besides our regular half parcel. I didn't stay but a minute because I wanted to get back to the combine and tell you guys."

"Charles," Clark said with authority, "you are probably correct about the parcels arriving. That is good news, for we are almost sure to get some extra food for Christmas; however, the rest of your rumor isn't worth a damn. When the general decides what the rations will be, the parcels will have been unloaded, counted, and stored in the warehouses. He will then have to obtain the approval of the Commandant. When this has been done, he will issue a joint statement to all compounds telling all of us at the same time. It will probably come in the night news report. The last part of your story is strictly a wild guess, and I think the originators of that rumor are off the beam."

"What do you really think, Clark?" Ed Wunn asked.

"Here is what I predict the general will do," Clark replied. "He will give us a half parcel per man of the Christmas food plus our regular half parcel of rations. This will give us a full parcel per man. He will do this Christmas week only, unless some more parcels arrive between now and New Year's. If some more parcels arrive, he will probably do the same thing during New Year's. If more parcels do not arrive, he will use the remaining six thousand Christmas parcels as a regular parcel issue."

After lunch, I went to the library with Gene and Ted Carson. Ted and I checked out two books, but Gene decided to spend the afternoon in the library.

That night I fell asleep reading.

The next thing I heard was McGee shouting: "Appell, first call for appell. Fall-out in five minutes."

When appell was over Pop-Eye stood by the north goal post and lectured Dumbkoff. Dumbkoff was slow on the count and Pop-Eye gave him hell. About twenty Kriegies were standing ten feet away watching in amusement. Bill Mott, standing in the center of the group, threw a large snowball and hit Pop-Eye on the back of the neck. Pop-Eye cursed a blue streak, and put the entire group in a brace.

"Which one threw the snowball?" he shouted. "The man that threw that snowball raise his hand."

All twenty men raised their hands.

"Very amusing. You lieutenants are always very amusing. When the war is over you won't have so much time for tricks."

The group all joined in with him. "The Fuehrer is taking you to Berlin to rebuild the city."

"Silence," he shouted. "The very next time I get hit with a snowball somebody is going to solitary. Is that clear?"

We all nodded.

"Dismissed," he shouted and joined Dumbkoff.

Back in the combine, we all laughed about Mott hitting Pop-

Eye with the snowball. Mott told us: "I was depressed. When I hit him with that snowball the veins protruded in his face. He got so mad he was slobbering. When he gets mad I always have a delightful sensation. I tingle all over. That will last me for at least a week."

After the dinner dishes had been washed, Dave Roberts, Ed Wunn, Matt Johnson and Peter Glass started a poker game. Pirtle made them move to the far end of the table, so he could set up the Monopoly game at the other end.

The poker game was still going on when the reader came in with the news. We all gathered around the table. There were no new developments in the war news. At the conclusion of the news there was a special directive from General Vandermann:

"Gentlemen: Today we received 16,500 Christmas parcels. I am making the announcement regarding parcels early in order that you may plan your Christmas accordingly. I signed the agreement with the Commandant this afternoon. You will be issued full parcels for the Christmas week starting Saturday December 23 and ending Saturday December 30. These parcels will be divided half and half. The remaining Christmas parcels will be issued in January during the regular distribution of rations. If and when more parcels should arrive, I will make another announcement at that time.

"We will be given freedom of the grounds Christmas Eve, Christmas Night, New Year's Eve, and New Year's Night. This afternoon I signed a parole for the entire camp with the Commandant. We will be allowed to go from barracks to barracks and stay up as late as we please. Any man who does not want his parole signed will give his name to the news reader after this message. Those who do not agree to the parole will report to the compound commander on these four nights. You will be locked in the kitchen after dark, and you will be guarded by the German guards. I think all of us will welcome the parole in return for the privileges granted us during Christmas. Let me make it clear that any man attempting escape who has agreed to the provisions of this parole will face American Military Court-Martial if his escape is successful. I will strongly recommend in such case that you be dishonorably discharged from the Army."

During the next few days, vast preparations were started for the holidays. The eating festivities were to start on Christmas Eve and continue through New Year's Day. Each combine elected three special cooks to prepare and plan the twenty-one meals from Christmas to New Year. Great plans were made for extra dishes such as cakes, cookies, pies, and candy. The Christmas parcels were comprised mostly of food. Among the most choice items were canned turkey, assorted nuts, fruit cake, canned fruit, chocolate, salt and pepper, and a box of pure sugar. There were also Christmas decorations around the inside of the parcel.

Friday morning, while each combine was drafting its elaborate plans, Captain McGee called me to his room. He and Captain Ferrell were stretched out on their beds, and McGee told me to make myself at home.

Sam McGee, twenty-eight and handsome, was of a very serious nature and seldom smiled. Bill Ferrell was good-natured, very punctual and efficient, and looked to be the same age as McGee.

I sat at the foot of Ferrell's bed and lit a cigarette.

"Simmons, we have been watching you, and we like the way you operate. Byron Clark recommended you, so we have decided to take you into Intelligence," McGee explained. "You will go to the library today and tomorrow to work on the camp record. You will also be given a certain part of the record to memorize. If we are evacuated from this camp, you will carry that part of the record with you. You will hide it on your person, and if at any time we are searched, you will be responsible for destroying it. When we are liberated, you will be expected to either produce your part of the record or dictate it from memory. You do not have to take this assignment if you don't want it. What do you think of the idea?"

"Well, Captain," I said, "I am certainly shocked and surprised. I am also flattered. How much will I have to memorize, and who will give me the documents if we are evacuated?"

"You will have to memorize several pages. It will all have to be done in the library. If we are evacuated, I will give you the documents."

"I will be glad to do my very best," I said.

"If we are evacuated, and if you are caught with the documents we are going to give you, you could be shot. Do you understand?"

"Yes, Captain," I answered, half scared to death.

As Ferrell and I left for the library, he said: "Don't take it so seriously. We are assigning parts of the record to two hundred men to carry. We may never even be evacuated. We are just getting ready, in case. Besides, the chances are they would never search us anyway."

We sat down at a table with Captain Randolph and Lieutenant David Bowling. Captain Randolph was second in charge of George, and he had assigned me my duties when I worked on George. I had never met David Bowling, but I remembered him very well. He was the daring chap who stole the bundle of maps from Rhoder's wagon.

Lieutenant David Bowling, aged twenty-two, was one of the best athletes in Sagan. He was five feet eleven, weighed one hundred seventy pounds, and he could walk across the parade grounds on his hands. His eyes were blue; his hair was brown and parted on the right side; his personality was excellent. He spoke German as well as Pop-Eye, and he could outsmart Rhoder any day of the week.

After I met Bowling, we each got a book from the shelves and moved to a table in the far corner of the room. Randolph started the conversation.

"Ferrell, you work on the record and re-check these documents. Let Simmons read his part of the record and start memorizing it. Bowling is here for memory work also. I want all of you back here after lunch, and I will want you here most of tomorrow. After that you will be checked by your block commander for progress."

"Randolph, how are my men in Combine F doing on George?" Ferrell inquired.

"They are doing fine. In fact, things are running smoother than at any time since George was started. Rhoder's wife is giving him so much hell about shortages and Christmas preparations at the German quarters, I haven't seen him in three days. We have a special parcel we have fixed for Rhoder's and Pop-Eye's wives. Maybe that will keep them out of our hair for a while. By the way, George is two hundred forty-nine feet long."

When Randolph left the library, Ferrell gave me four typed pieces of paper. I opened them and started reading. They covered copious details of several atrocities committed against individual American prisoners. There was also a report on inadequate medical facilities and of medical treatment denied prisoners in critical need. Two men who had died from lack of medical treatment were listed, and their place of burial was given.

I read the entire report over three times before lunch. After lunch I returned to the library, and started memorizing the atrocities committed, and the name of each man involved. I memorized all names, places, and dates in the report. I then closed the report and started telling it to myself in story form.

Saturday morning I missed the inspection and reported to the library. I had forgotten some of the names and dates, but I had the entire story in proper order with each location described correctly in its proper place. By Saturday evening, I had repeated the entire report several times without error. When I was dismissed Saturday evening, I returned to the combine, tired.

XVI

The Great Bash

THE PREPARATIONS FOR CHRISTMAS WERE GOING FULL SCALE AHEAD. Everybody had the Christmas spirit, and ninety per cent of the men were working on various projects for the Christmas celebration.

Bill Mott had been elected head cook in charge of the Christmas week meals. Bob Rhinehart and Dave Roberts were elected his assistant cooks. All evening meals were to be eaten formally. Pirtle was in charge of decorating the table for each meal, and he had made a list of things for each member of the combine to make. Clark was elected to decorate the combine, and was given Charles Swanson, Glass and Johnson as his assistants. Ted Carson, Ed Wunn, and I were elected to the entertainment committee. We were to provide and arrange for special entertainment for the combine during the Christmas week, and it was also our duty to keep the combine posted on all worthwhile entertainment in the compound. Gene Coletti was elected chief K.P. He was responsible for the washing of all dirty dishes and for keeping the combine in tip-top shape during the week. Excluding the three cooks, he could choose two men per day to help him. It was agreed that all members would do their part in keeping the combine clean.

Plans for compound entertainment were on an equally large scale. Christmas entertainment festivities were to start on Thursday, December 22, with a special sing-song in the theater. The band and choir were in charge of the program. A Christmas play was scheduled for Friday night. Saturday night was Christmas Eve, and a Christmas Pageant was scheduled which included the band, choir, comedians, actors, and others. At dark on Christmas Eve, carols were to be sung in every barracks of the compound, with a member of the choir leading the singing in each block. Christmas Eve was to end with midnight church services; Christmas Day would be highlighted by three special church services; Christmas evening the band would play a special program of popular music, and close with several songs to be sung by the entire audience. Christmas morning General Vandermann would visit every combine in each compound to select the best-looking cake. The combine with the best-looking cake in each compound would be awarded an extra Christmas parcel. During Christmas week the band would give a three-hour show in each barracks of the compound. The show was called "Night Club," and there was to be music, fun, and dancing for everybody.

As Christmas grew near, plans grew more elaborate and the Christmas spirit grew stronger. By December 22, every barracks in the compound had been decorated. Some decorations were home-made, and others came from the parcels. There were two or three presents in each parcel, such as smoking tobacco, pipes, and other small gifts. Each of these gifts was wrapped and names were drawn from each combine. Every man was to have a Christmas present. All preparations for meals were complete. All Intelligence's projects were shut down, with the exception of Dave. Ten thousand prisoners at Sagan settled down to celebrate the Christmas holidays.

On Friday morning, December 23, Mott gave every member of the combine a haircut. It didn't require too much time, as he trimmed most of us once a week to keep us looking tip-top.

Friday morning I was called to McGee's room. We sat on his bed, and I repeated the entire four-page report of the camp record. I went over it twice each day, because that was the only way I could remember all of the names, dates, and details.

Friday night Combine C attended the second showing of the Christmas play which was the Christmas story of the birth of Christ. Seven hundred men filled the auditorium for the second showing. The acting was excellent; and the choir remained behind the stage curtains, supporting the play with Christmas songs. The music was beautiful, and the play was heartwarming. After the program was over, the band and choir led the entire audience in the singing of carols.

Saturday morning the parcels were issued with much excitement. Rhinehart let every member of the combine help open them, and the table was loaded high with their contents. We all watched like children looking at new toys on Christmas morning. Pirtle insisted that everything should be put away before we destroyed part of it. Everybody watched Rhinehart and Mott stack the precious items on the shelves.

Saturday evening at dusk the band and choir moved across the parade grounds. Many men came outside the barracks while others stayed inside. The moon had come out, and as its light spread over the camp, the snow looked like tiny diamonds sparkling in its glow. As the Christmas carols started, thousands of male voices rang throughout the barracks, across the open grounds, and into the open air. I know the city of Sagan heard us that night, for the music and thousands of voices echoed across the skies. The German guards on duty lowered their rifles to the ground, and listened in silence. The guards in the watch towers came outside and removed their hats. The last carol to be sung was "Silent Night." As every prisoner joined in to sing, there was a warmth in that cold, freezing air that only God could make. It was a strange feeling. All fear was swept away. A feeling of peace engulfed every man. When the music stopped, we moved slowly and silently inside.

Saturday night we dressed in our Sunday best for the Christmas pageant. There were two shows. One was from 6:30 until 9:00 p.m. and the other was from 9:00 until 11:30 p.m.

Pirtle paced the floor until Clark returned from the raffle. Clark was smiling. "He did it again," Pirtle hollered. Clark had drawn the first show, seats in the left hand section.

During the two and one-half hours of the Pageant, two one-act plays were presented about Christmas. The band played popular music for half an hour, and the choir sang all of the carols. Many jokes were told. There were clowns and entertainers of all types. The show closed with the entire cast coming on stage and singing "Silent Night."

When the last show was over, Catholic Midnight Mass followed in the auditorium. We didn't even try to go, for the auditorium was packed to capacity.

Early Sunday morning General Vandermann's Christmas message arrived by special messenger.

"My Fellow Prisoners: Let each of us thank God for this beautiful Christmas morning. We should especially remember Him today for the hundreds of blessings He has shed upon us throughout the year. As we now approach the end of a long, desperate, and trying year of war and prison life, let us be thankful. Grateful that we are alive and in good health, but most of all grateful that we have God as our guiding light to watch over us and protect us. Let us thank God for all He has done for us, and let us ask Him to have mercy on those who are less fortunate than we. Let us ask Him to bless all people, and fill their hearts with love and happiness. Let us ask Him to bring this war to a rapid conclusion, so that misery and suffering throughout the world may come to an end. Let us ask Him to give us strength and courage to carry on as we have done in the past; so that when the final day and hour comes, we may rise and walk with His blessing, through these prison gates.

"Merry Christmas and a Happy New Year to each and every one of you. God bless you and keep you brave and strong in the trying days yet to come. May you remember these words: enduring courage, strong hearts, prayers, and faith in God will overcome any force on this earth."

The Christmas morning church services were as beautiful as any I had ever attended at Sagan. The choir, the scripture readings, the songs and prayers touched each man deeply. The entire sermon was centered around General Vandermann's message of faith and hope. At the conclusion of the service, the auditorium was emptied in silence.

When we reached the combine, the cooks had dinner on the stove. In a few minutes Mott and Rhinehart came through the door with pots and pans of every description. We all stayed on our beds and smoked and watched in silence. When the signal was given, we gathered about the table.

Clark turned off the light, and Pirtle lit five candles spaced about the table. The room was semi-dark, but the candlelight gave off a pleasant glow which added a touch of beauty to the table. The large table had been covered with a white sheet. The benches and chairs were decorated with red and green streamers taken from the Christmas parcels. In front of each plate was a Christmas hat, designed and painted by Mott. To the right of each plate was a small can of candy and nuts equally divided from the parcels. Clark had donated five packs of assorted cigarettes, and placed them in a can in the center of the table, so each man could smoke whenever he desired. Clark asked me to return thanks. When all heads were bowed, I said grace.

The meal consisted of turkey and dressing, fluffy white mashed potatoes with thick brown gravy, creamed peas, carrot and cabbage salad, toasted cheese sandwiches, hot coffee and buttered rolls. Pirtle put the jar of coffee on the table, and let each man strengthen his own. The dessert was three slices of chocolate cake for each man, with thick chocolate icing.

Each of us ate slowly, enjoying every delicious bite. We ate and continued to eat until it was impossible to swallow another mouthful. The real desire among all of us was to be full and in want of nothing. We believed that by eating enough we could drive away the pang of starvation. This was a meal we had dreamed about, and we stayed at the table until no one could eat another bite. There was food on the table when the meal was concluded, and it was left there for anyone who might wish to nibble later in the day. The cake was stored on the shelves for future consumption. It was necessary for the K.P.'s to rest over an hour before they could start doing the dishes. There were contented groans everywhere. For once, every man in the barracks took a good nap with his stomach full. It was late evening when Combine C finally came to life.

Byron Clark was sitting at the table drinking coffee with Gene, when he told the group: "I have dreamed of this meal for eight long months. I never knew before, in my entire life, that food could taste so good. We have all eaten turkey before, but today I looked at that turkey and ate it as if I had never seen or tasted it before. Look at Gene. He has drunk hundreds of cups of coffee in the past, but never a cup with such an aroma as the five cups he has consumed today."

Tom Pirtle joined us and said: "I wonder sometimes if we will ever be able to explain all of this to the folks back home. Yet, how could we expect them to understand, unless they, too, could see and experience what we have seen and experienced. In every major crisis of prison life God always seems to appear here at Sagan. We all know He is here. Maybe He stays His distance, but we always know what He is saying. 'Have faith and keep going, and I will guide you.' I have heard it, just as if He poked me in the back and said it to me personally."

Clark said: "Maybe in time we will find the answers."

"Let me point out something that has always been strange to me," Rhinehart said, as he came to the table and poured a can of coffee. "We have heard many rumors about how General Vandermann was assigned to this camp. The best one seems to be that he flew over this area in a bomber and bailed out. There are many American, British, and French generals who are prisoners of the Germans but all of them live in several manor houses or castles. All of the Allied generals but one, and that one is General Vandermann who lives at Sagan. Isn't it strange to you that General Vandermann is German, though an American citizen? Isn't it also strange that he was Military Attaché to the American Ambassador in Berlin before the war? General Vandermann knows the top bigwigs of Nazi Germany. Look at the big wheels who have visited him here. He speaks several foreign languages, including the languages of our enemies and allies alike. Look at our life here at Sagan. Where else in Germany do the Germans abide by the Geneva Convention regarding prisoners of war, to the letter of the law? Our quarters, our recreation, our privileges, our food, Red Coss supervisors entering and leaving the camp as they please—and the overall operation of this prison is most unusual. General Vandermann is constantly negotiating with the Commandant about the slightest thing that he does not consider proper. Vandermann is well-known in Germany, and, of course, we all know the Commandant is scared to death of him. We are living a most unusual prison life. This Christmas is a good example. We wouldn't find such conditions anywhere else in Germany. We all know that Vandermann made all of this possible. Look at our future. We feel safe with him. Why? Because we know he will lead us and be with us wherever we may go. Now the point is this. Who sent him here? Was it planned and if so, by whom? Did he just happen to come here by accident, and did all of these things just happen by accident?"

Again Clark answered: "Only time will tell."

Charles Swanson was serious for the first time since I had been in Sagan. He spoke softly. "In our prayers we have forgotten someone. Chaplain Daniel always mentions it in church, but I think our combine should also do it. Everything we have before us has come from the American Red Cross. This meal today, and all of the food we have eaten, has been furnished and delivered by the Red Cross. They have furnished our clothes, the library, band instruments, athletic equipment, blankets, our dishes, and just about everything we possess. Those parcels also give us our soap, candy, cigarettes and matches. Every combine game came from the Red Cross. All in all, we would be dead had we not been blessed with the American Red Cross. We actually owe our very lives to them."

While the conversation continued, Mott and Rhinehart warmed the turkey, potatoes and leftovers for supper. Gene went to the kitchen for hot water, and an hour later we ate a small snack. Several of us could eat no more than three or four bites.

XVII

The Escape

ON MONDAY, DECEMBER 26, THE PAROLES SIGNED BY GENERAL Vandermann expired, to be renewed on New Year's Eve. Freedom of the grounds ended, the full German guard returned to duty, and the normal schedule was resumed.

The Sagan band started touring each barracks for its "Night Club" performance of music, dancing, and fun. The "bash," with mountains of food to be eaten, continued according to plan.

Pop-Eye attended appell, and wished everybody a belated Merry Christmas. He smiled and laughed as he checked each block, until Dumbkoff fouled up the count in block seven. He tried to restrain himself, but Pop-Eye must have decided that Dumbkoff was getting cocky. He called Dumbkoff and Shisenkoff before us, and placed Shisenkoff in charge of the front ranks counting noses. Shisenkoff was thrilled, and Dumbkoff suffered great humiliation in being demoted to the rear ranks to count Kriegie butts.

Breakfast was served at 8:30 a.m. It consisted of a bowl of barley covered with milk, two cookies, two pieces of bread, a slice of baked spam, and coffee according to our own desires. Truly a wonderful breakfast enjoyed by all members of the combine.

"Oh, hell, here comes Swanson back from the latrine," said Pirtle.

Swanson tripped over the bench by the stove, and fell against the table between Pirtle and Mott.

"I knew it, I knew it. It happens every time," Pirtle moaned. "Dammit to hell, I am through gambling on Monopoly, poker, or bridge. From this day forward, I am going to bet all of my cigarettes on Swanson hitting something or tearing something up in this combine. This way I will win every time."

"There's going to be a mail call in about fifteen minutes," Swanson announced. Everybody leaped from their chairs and beds.

"How do you know?" we all asked.

"I was in the latrine and . . ."

"Here we go again," Pirtle shouted.

"Well, I was in the latrine," Swanson repeated, "and Burns from block six was siting next to me. Anyway, McDonald came in and told Burns a lot of mail had arrived, and they were sorting it in the auditorium."

"Swanson, if this is a rumor you won't have to worry about our liberation. You won't live that long," Pirtle told him.

Most of us decided it was time for another lap or two around the track. We dressed in overcoats and left the barracks. When we got to the track, there was hardly room to walk. Everybody was trying to deflate their expanded stomachs. On our third lap around, Swanson hollered: "There goes McGee with our mail!" Everybody broke into a run. McGee got in the center of the hall and shouted: "Mail call."

That was our first mail call in two weeks, and everybody sweated it out until the last name was called. Every member of my combine got mail, except Gene, Dave Roberts and me. We stayed in the combine, and watched the others read their letters.

Pirtle was the first to pass his mail. "Here, Gene," he said. "You want to read a letter from home?"

Most of the fellows passed their letters around, letting all of us read them. It was wonderful to read a letter from any mother or girl friend, just so long as it was from the United States. I read a dozen or more, and they all said just about what one's own mother or girl friend would have said. Some fellows stayed in the sack for several hours re-reading every line over and over. I couldn't figure out how our Christmas could have been better.

Each week the goons gave us three single sheets of paper that folded into an envelope. I had written fifteen letters to my family, several girls, and my closest friends. When you finished writing, you would fold the outer edges inward and seal the sheet. The outside surface of your letter had a place for name, street, city, and state. The goons would censor the mail and stamp it for mailing. After a while I would start getting mail from home.

Monday evening it was announced that the Sagan band would play in block eight of our barracks. Block eight moved the tiers in several combines against the walls to make room for the band and dance floor. Our block joined block eight, and at seven that evening the band arrived. There were many dancers with the band, each dressed as a girl. We had the time of our lives. They played popular music for three hours, and the "girls" danced with everybody. I never got to dance, but those who did easily imagined they were dancing with real girls.

Tuesday morning I was called to McGee's room.

"Good morning, Ken. Let's go over your part of the record," he said. I repeated it again, but owing to the excitement over Christmas I had failed to review it, and I missed two places and a date. McGee gave me a good lecture, and I knew from that day I would never forget another word of that report.

That night for supper we ate another delicious meal. An extra vessel was filled with hot water and placed on top of the stove before we ate. I wondered why all the hot water; at 9:00 p.m. I found out.

Pirtle and other coffee fuehrers like him were mixing coffee all over the compound. When we were seated at the table, Ferrell came into the combine and had a cup with us. When he sat down at our table he said in a calm voice:

"There will be an attempted escape tonight after lights out. One man is going to try to get out. This escape has been planned by Intelligence for a long time. Our man has been drilled and prepared. The plan of escape will be revealed to you after the goons find out about it. This man has been equipped with German civilian clothes, wire cutters, compass, maps, train tickets, train schedules, several hundred dollars in German marks, food, German identification card, and all other necessities.

"Our man has been sleeping in the snow for several nights under one of the barracks to get in condition. We have been feeding him triple rations and he is in excellent physical condition. He is outside now, awaiting the zero hour.

"He is carrying a very valuable report from General Vandermann to Allied Headquarters. If he gets through, it will be of utmost importance to all of us. If he is caught outside the camp, he will destroy the report. It is hidden on his person in a safe place. If he is caught escaping inside the camp, of course they will probably shoot him.

"We will all act as if we were totally surprised whether he gets through or not. Everyone will act very teed off about the whole affair. Now if he gets through, the goons will eventually find the holes in the fences. When they find the holes, we will know. The sirens will start blowing, and the lights will be turned on all over the camp. There will be more guards in this compound than you ever dreamed possible.

"Your job will be to stall and delay the goons the longest possible time. Act dumb, raise hell, and take as long as you can to dress and fall outside. Every minute we delay them will be an added minute for our man to make his escape good.

"Remember that they can't start sending out descriptions, radio reports, or patrols to hunt him until they first know how many men escaped, and who they are. They will try to find out the number missing by an appell. We have men in every block prepared to foul up the count. If they ever get the appell correctly, they will know only one man escaped, but they won't know which man. They will then have to have a picture parade to identify us one at a time, in order to tell which man is missing out of the two thousand men in our compound. The picture parade will involve hours of work and identification. This will give our man a good chance to make it. If they ever get the appell correct, they will broadcast over the radio the number of men who have escaped and alert many organizations to start looking. We must do everything possible to keep that appell fouled up all night. Only the men that have been designated will move in the ranks of each block. They will have plenty of goon guards, and we don't want anyone to get shot.

"This may be an all-night affair, but we must give our man several hours head start. If we have good luck, he will be on his way out of Germany before they know whom they are looking for.

"Remember, act dumb as hell and mad about the whole affair. I can't wait to see Rhoder's face."

Just before lights out we had another cup of hot coffee as we undressed. We knew the escapee would start about thirty minutes after lights out. We got in bed, but none of us tried to sleep. We could picture our man cutting through the fences. We listened for shots, but none came. We strained our ears to hear the chatter of excited German guards, but we heard nothing. The spotlights from the towers continued to roam the grounds and sides of our barracks. Every now and then we could see the tower lights moving past our blackout shutters. Clark quietly called the time every half hour from his bed. Just a few minutes after Clark called midnight, the goons made their discovery. We all sighed with relief, for we knew our man had gotten out of the camp.

The goons changed the guard at midnight, and the new ground guards made their rounds with German police dogs checking the outside fences.

At first we heard one goon shouting in German. We heard another goon answer. Then we heard several guards shouting to the tower guard. Then we heard the sirens. Two long blasts repeated several times, signifying an escape. The blasts could be heard all over the area. Goons were rushing and shouting outside. Some goon was issuing commands. He was shouting so loud we could hear him in our combine. Three minutes later the lights came on, and the compound was lit up like a Christmas tree. Many of us pretended to be waking up, while others pretended to sleep. The barracks doors swung open from both ends and guards stormed into the barracks. Ferrell wasn't kidding about the number of guards. Eight guards entered each end of our barracks and started shouting at the top of their voices:

" *'Raus! 'Raus!* (Out! Out!)"

They came down each end of the hall armed with rifles and pistols, and spaced themselves about ten feet apart and shouted at all of us.

In a few minutes Pop-Eye came storming through the barracks. "Oust, oust," he commanded as he checked each combine. He shoved several men against their beds or the wall, and told them to dress immediately or they would stand outside in their pajamas. His face was red as a beet, and his temper had reached the danger point.

McGee asked him what was going on, but he was so mad he couldn't make sense. He put McGee in a brace, and shouted in his face.

"Liar, idiot, luft gangster. Somebody is going to die for this, and it isn't going to be me. Do you understand, liar, idiot. Oust! Oust!" he shouted at McGee as he pointed his hand toward the barracks door. Without another word he stormed out of the barracks. The guards remained at their stations, jabbing rifles and hurrying us outside.

When we started coming out of the barracks, goon guards were stationed at each exit with terrifying German police dogs. Every Kriegie in the compound was turned out for the appell, including the general. The commandant and his entire staff were by the north goal post, and the look on the commandant's face was confirmation of real danger to us.

Every block was called to attention and the counting started. You could hear the German guards in every block counting "ein, zwei, drei, vier," as they proceeded down the lines. Our lines were three columns deep. When the first eight or ten columns had been counted, the end man in the front column would take one step to his left. The man behind him in the middle column would take one step forward, replacing him. All men to the right of the man in the second column would take one step to the left, filling in the gap. The right end rear rank man would take one step forward to fill the vacancy in the middle column. This maneuver could be worked any number of different ways from different ends and different columns. The count could be thrown off short several men per block when two guards tried to count the entire block. There were twenty blocks in the compound, and it took forty guards to make a regular count.

Intelligence had instructed the first two counts to be way over the number of men on roll in the Center Compound. The roll of the Center Compound was 2,006. As each goon in charge of each block count reported to Pop-Eye, he listed the block number and the number of men present.

When Pop-Eye got the total he looked at his roll and re-added the count. The count totaled 2,042. He reported to the commandant that there were 36 too many men, and the commandant cursed him out. He cursed all the German guards, too, and the count was ordered retaken.

On the second count every time a counter saw a Kriegie move, he would start over. The German counters cursed and shook several Kriegies. Shisenkoff and Dumbkoff counted our block four times during the second count. The second count totaled 2,012. That was still six more men than the total roster of the Center Compound. Pop-Eye's veins were protruding in his face as he once again reported to the commandant. Rhoder had just joined the commandant, and the veins in his face were also protruding as he kept looking at his watch. The time was 1:16 a.m. The commandant issued an order to Rhoder. Rhoder and the commandant's chauffeur drove off in the commandant's car. A small platform was brought to the north goal post by two guards. It was placed there for the commandant to stand on to avoid getting his boots wet in the snow.

We stood at attention for fifteen minutes before Rhoder and the chauffeur returned. Rhoder quickly reported to the commandant. They talked for several minutes, and then the guards really started coming. They were moving on the double as they came through the main gate of the compound. We estimated 150 additional guards plus Rhoder's men. They were carrying rifles, pistols, submachine guns, battery-operated spotlights, and billy clubs.

The commandant called General Vandermann to the north goal post, and spoke to him several minutes. Rhoder and Pop-Eye placed eight additional guards in front of and behind each block. They unslung rifles and took up firing positions. The spotlights were turned on each block. We were informed that any man moving about would be flogged or shot.

The third count was taken and re-checked. It was impossible for anyone to move, and this time the counters gave a correct report to Pop-Eye. He totaled it up, and it came out 2,005. While this was going on, two large flat-bed trucks came into the compound and stopped in back of the goal post. Tables and chairs were erected on the bed of each truck and lanterns were placed on each table. Six large file boxes were placed on the six tables. We knew what was coming next.

These boxes carried our identification cards and prison numbers. Each card had our name, rank, serial number, prison number, fingerprints, and picture.

After Pop-Eye reported to the commandant that one man had escaped, the Germans really swung into action. The commandant's aide and chauffeur jumped in his car and started for Camp Headquarters. We knew they would report one American prisoner escaped from Sagan, so it could be broadcast to the authorities. The report probably said that identification would follow.

When the picture parade started, it was 2.08 a.m.

The picture parade was the identification of each individual prisoner. Block one was marched with twenty guards to the center of the two trucks. Spotlights were placed on the truck and shined in the faces of the Kriegies of block one. Pop-Eye and Rhoder started at opposite ends of the front rank removing the prison tags from the neck of each Kriegie. They would call off the prison number. All cards were filed in numerical order. The ferrets at the card boxes would pull the card by number and pass it to Rhoder and Pop-Eye. They would check the picture and the Kriegie to be sure they were the same. As each man was identified, his card and picture were re-filed in another box. When the identification of front ranks of block one was completed by Pop-Eye and Rhoder, five guards marched the front rank of block one to their barracks. The guards with German police dogs prevented anyone from leaving the barracks. The second rank of block one was then checked and so the picture parade continued on into the night. It was 3:50 a.m. when they finished with block seven. We were almost frozen when we got back in the barracks. All lights stayed on, and we stayed up to hear the news.

When they finished with block eighteen it was 6:02 a.m. One man was unaccounted for, but the commandant wasn't taking any chances. He ordered the picture parade to continue. A few minutes before seven every American prisoner was locked inside, and only one picture and identification card remained in the boxes—First Lieutenant David Bowling, block eighteen, center compound.

McGee had already told us that it was David Bowling. I remembered Bowling stealing those maps after the search. I remembered him with me at the library when I was learning my part of the camp record. I also remembered Captain Randolph saying that Bowling was there for memory work too. But I never dreamed he was memorizing a message that he would try to carry to Allied Headquarters.

The guards returned to their respective posts. The commandant left in his car and Pop-Eye and Rhoder examined the holes in the fences and the path of escape. A complete description would be issued immediately to all authorities describing Bowling. There was one pleasing aspect to all of us. It was seven a.m. and Bowling had escaped at ten-thirty p.m. the night before. He had had eight hours with no one trying to find him. By now we felt that he was a long way from Sagan; by making proper train connections he could be halfway across Germany.

McGee stationed himself in the center of the hall, and told us the story. Dave Bowling had been under barracks eighteen when we were locked in for the night. He was to start his escape at 10:30. The snow was nearly two feet deep. Bowling had his bundle tied to his belt. He started by lying flat on the ground under the barracks. He had a white sheet which he pulled over his entire body. He stretched the sheet over his head and out from him with both arms extended. When the tower lights started coming his way he would lie face down in the snow with the sheet covering his body. From the watch towers it all looked like snow. As soon as the light moved past him, he crawled with the sheet over his head toward the first fence.

At the first fence he waited for the tower lights to pass by. He then used his wire cutters, cutting a hole large enough to crawl through. When the tower lights passed him the second time he went through the fence, sheet and all. At the second fence there were no tower lights, but only guards trooping the fences. The guards had been timed and they made their rounds every fifteen minutes. As soon as the guard had passed, Bowling moved to the fence, cut his hole, crawled through, wired the fence back, and crawled with his sheet to the woods. From the woods he changed clothes, buried his prison clothes and sheet, and took off.

It sounded simple, but would have been impossible without the snow. Bowling had practiced, trained, and timed his operation hundreds of times. He had lain under his barracks at night, timing the spotlights and timing the guards making their rounds. He had practiced cutting wire until he was an expert. All in all, it was a brilliant, well-planned escape.

"Tallyho" came the shout through the barracks.

Swanson came rushing into the combine and jumped on his

bed. He said: "It's Rhoder, and he doesn't act as if he's very happy."

He stopped in three combines before he arrived at Combine C. Clark offered him a cup of coffee, and he accepted.

"Well, my dear lieutenants, you have gotten us all in hot water. You couldn't let well enough alone. You just had to stage an escape, so you would get the commandant's name in the papers; and probably a very bad report before some general in Berlin. The commandant is very unhappy, and that makes me very unhappy. If he gets into trouble, then I will get into serious trouble. I have tried to cooperate with you, but you must play your little games. Very well. We are going to start putting a few of you in solitary, and there will be a few other surprises in store for you. General Vandermann doesn't run this prison. He is our prisoner and we are likely to put him in solitary. Some of you think the commandant is afraid of your general."

Clark interrupted. "We didn't say we thought the commandant was afraid of the general."

"No, Lieutenant, but you think it just the same."

"We were in bed asleep. We didn't know a thing about the escape. Just because one man tries to escape, how can you expect all of us to know about it?"

Rhoder frowned, but kept his temper. "I admit that I am not as smart as the Fuehrer, but I am no fool. You all knew about the escape. You planned it, timed it, and acted it out just like all of your tricks. Someone is going to pay for this; and if Berlin makes a complaint to the commandant, all of you are going to suffer and suffer."

He got up from our table and left.

On New Year's Eve the parole was not granted because of Bowling's escape. We had parties in our barracks, and a late New Year's supper. We stayed up until midnight, and watched the year 1945 come in. The Germans left the lights on until 1:00 a.m., and Dave was read shortly after twelve. The war news on the Eastern Front caused wild excitement.

The Russians had reached the German border, and were moving up the largest reinforcements of the war. Several Russian armies were being moved to the area, besides the armies holding the lines, for an all-out assault against all of Eastern

Germany. The report said the Russians were moving thousands of planes and tanks and dozens of armored divisions for the final all-out assault. The estimated strength of Russian forces on the Eastern front was nearly four million men.

We knew now that the war could not last much longer. The American, British, and French armies were along the western borders of Germany, and the Russian armies were at the eastern borders of Germany. Advances on either front meant advances into Germany proper. This news gave us great hope for the coming year, and men like Pirtle became excited for the first time in many months.

On January 2, 1945 we returned to normal operations of the camp, and all Intelligence projects were resumed.

That morning when we made the shower parade, I recalled my first trip to the showers, the day I came to Sagan. The German sergeant had said we would learn how to take showers, and he was correct. As older prisoners, we would gather five men under each water spout before the water came on. When the water was turned on, each man would quickly wet his body and step out from under the water to soap from head to toe. When the fifth man was soaping, the first man was washing off. Every man was clean and in the dressing room drying when the water went off.

Most of us took washrag baths each day, sometimes between breakfast and dinner. These baths were taken with a pan of cold water, washrag and soap. It wasn't like a hot shower, but it did help.

Tuesday afternoon David Bowling was returned by the S.S. to the Center Compound. He had been captured near the front lines on the western sector by German troops. Bowling had made most of his journey by civilian trains, and by riding in boxcars. It was during the last leg of his journey on foot that he was questioned and discovered. He had destroyed his secret message his first night in jail. The Germans had searched him thoroughly, but never found it. He had been carrying it in a removable heel of his G.I. shoe.

The commandant sentenced him to ten days of solitary confinement, and the incident was closed. It was the closest any man ever came to making his escape good while I was in Sagan.

During the remainder of the first week in January, 1945, the main events of the news centered around the tremendous Russian build-up on the Eastern Front.

Hope and confidence grew among us, and we really began to believe that the war was in its final stages. Many older prisoners tried to remain calm, but the large majority of my fellow Kriegies became optimistic, and a feeling of victory and liberation spread over the camp.

At the same time our speculations about the future often took a dark turn. Pirtle summed up our conjectures one night.

"Knowing the Germans as we do, it won't be that easy. I have been thinking about the many possibilities during the entire week, and I have come up with several answers. Some of them are gloomy and only possibilities, but I like to face the situation squarely and look at the facts. The Russians could advance so fast that the Germans wouldn't have time to evacuate the camp. Moving ten thousand prisoners of war is no easy task. If this should happen, General Vandermann might talk the commandant into surrendering. If he can't do this, he still might talk him into leaving us and evacuating his own personnel.

"Then we still have another angle to face. This is the largest American prison in Germany. High officials in Germany, such as fanatics like Himmler, could order us to be put to death. We must remember what Himmler did in the British compound over that tunnel. He had fifty men picked at random and shot. He is capable of doing anything. If the Gestapo takes over this camp, the commandant will be powerless and so will General Vandermann. The only other angle I can see is a possible evacuation before the Russians arrive. If this happens, we will still be in great danger. We will have to be moved a long way. The weather is terrible. We are in poor health, to say the least. We would be evacuated in the midst of battle, and a march of this type would cause many casualties."

"You're right, Tom," said someone I don't remember. "Three or four days of marching would put us in bad physical condition. From now until the moment of crisis will be a bad time. We must all work to keep up morale and keep rumors down. Rumors can paralyze morale, and any panic among us could mean our own self-destruction."

XVIII

The Russians Draw Near

TENSION MOUNTED AMONG THE KRIEGIES OF SAGAN AS TEN thousand American prisoners of war waited for the Russian attack on the eastern front. News continued to pour in of tremendous Russian reinforcements.

During the noon hour on Thursday (January 12) Captain McGee put guards on the barracks doors and climbed on a table in the center of block seven. Most of us stayed in our combines so as not to attract the attention of the outside goon guards.

"At ease," McGee shouted. "This morning at dawn, three Russian armies launched an all-out attack along the four-hundred-mile front of the German-Polish border. A tremendous battle is now raging between Russian and German forces. This news report was taken at 12 noon; Beuthen and Gleiwitz, Germany, have fallen to the Russians. According to this information, the Russians are 165 miles from Sagan. That's it."

We restrained ourselves from shouting, but we were wild with excitement. The long-awaited Russian attack had come at last; and now we could start counting the miles.

Nearly everybody started placing bets on the advances that would be made each day. Some men bet on the day Breslau would fall. Pirtle made three small bets on the day Sagan would fall.

On Friday night the news continued to be excellent. The Russians had won their first major battle, and the German army was in full retreat. Falkenberg, Gottesberg, and Waldenberg had all fallen. The Russians were not over 106 miles from Sagan.

The Germans knew that we were getting the news, but they

didn't seem to care. Busy listening to it themselves, they didn't bother with us. The German guards were afraid of the Russians, and each Russian advance increased their uneasiness and fear.

On Wednesday night, January 18, Dave was read an hour earlier. The table was crowded and no one bothered to deal the cards or open books. "Today at 1:00 P.M. Breslau fell to the Russians. The latest report tonight said the Russians are advancing on Liegnitz. From the present news reports the Russian army is now 85 miles from Sagan. The most important item in the news was the liberation of eight thousand American Air Force enlisted men in the Breslau prison. The Russians surrounded the prison, and the German commandant surrendered without a fight. The news reported high American military authorities as saying that all liberated prisoners of war would be flown within a few days to special hospitals in France.

"Here is a special message from General Vandermann: 'In the last few days various false rumors have been reported to me that I consider harmful to the welfare of this camp. Each of these rumors was started by some Kriegie in this camp. I want all rumors stopped at once. I shall enforce severe penalties against anyone who starts one of these rumors. The news we are giving you is accurate, and anything else you hear should be disregarded.'"

For the next three days, in spite of the general's order, rumors started from every latrine and spread in every area of the camp. Every man in camp began to pick the day and hour that Sagan would fall. It was rumored that the Germans were going to evacuate the camp. Another rumor stated that they would make a last-ditch fight. Someone started a rumor that the Germans were preparing to evacuate us. It was asserted that we would be held as hostages and shot if the Russians tried to take Sagan. It was rumored that the Germans could kill prisoners of war if it became impossible to evacuate them. Someone even started a rumor that the goons were converting the shower houses into gas chambers, and that we would be gassed if they could not evacuate us.

As the Russians drew nearer, the rumors became wilder and our future became more uncertain. Morale began to drop, as fear and the reality of death mounted in our minds.

General Vandermann launched a full-scale attack against rumors. He sent message after message each night showing how fantastic most of the rumors were. He sent members of Intelligence to every barracks to lecture on rumors and to show how they were destroying morale and camp organization. He stressed teamwork, unity, and courage.

In spite of all the general's efforts the rumors continued to spread and grow. Morale dropped to a low ebb, and all of us became lax about our duties.

On January 22 General Vandermann ordered every compound commander to prepare the camp for possible evacuation. He told each commander that we were to be kept so busy there would be little time for wild talk. Each compound commander and his staff organized their plan for possible evacuation, and sent out directives to all block commanders. Following this, each block was given detailed lectures, instructions about the evacuation, and rehearsals were scheduled. Each man was required to walk the perimeter track at least four times a day. Men were stationed on the perimeter track, and we were checked by roster to make sure we made the required number of laps. Lists were prepared on all items to be taken in the event of an evacuation, and we were instructed on how to roll and re-roll packs. We were to make our packs out of blankets, and we drilled for hours rolling and re-rolling packs with all items in the pack. After we learned how to roll packs, we had to take them outside and practice rolling them in the snow. Marching schedules were organized in each combine. The food packs of the combine would be rotated to each member of the combine for a twelve-hour period on the march. Suggested lists of clothes to be worn were posted on each block bulletin board. By January 24 the entire camp was prepared to march if the evacuation was ordered.

On the night of the 24th the news was reported with many new developments. Goldberg, Lauban, and Liegnitz had fallen to the Russians. The German army had stopped the Russians on only two occasions, but they were now reported in full retreat. These latest reports left the Russians 39 miles from Sagan. New rumors started up.

General Vandermann immediately started a second plan to be organized by compounds, in the event we were not evacuated. This plan was to be used if the Germans tried to hold the camp against the Russians, or if they tried to carry out any form of mass execution against us.

Groups were organized by blocks and by combines to storm every sentry tower in the compound, if mass executions were tried. In addition to this, ten men were selected and assigned to kill each German guard. Knives and weapons were issued to the leaders of each organized group. The plan was rehearsed and timed. If the Russians attacked from the outside, this attack would be launched immediately from the inside.

On the night of January 25, the news placed Russian troops 28 miles from Sagan.

We knew the Germans could have evacuated us days ago. They had delayed and continued to delay although they had hourly reports on the location of the Russian army. Now that the Russians were only 28 miles away, most of us felt it would be impossible for such an evacuation to occur. Some German guards, who were probably scared themselves, had actually told several prisoners that if we could not be evacuated there was a good chance we would be shot. When this story got over the camp, men became hysterical and discipline crumbled. Many men lost faith in everything. To top it off, the goons were becoming extremely nervous and several times a day or night a trigger-happy guard would fire his rifle.

On January 27 at eight forty-five A.M. we crowded into the auditorium. Most of the men in my combine had to stand. All doors were left open, and windows were raised. By nine o'clock lines had formed three deep from each door and window, and the entire compound gathered in and around the auditorium to hear the general. It was extremely cold inside and outside, but no one seemed to notice it.

Promptly General Vandermann approached the main entrance and the word spread to every man:

"The general. The general is coming."

Colonel Spivey, standing on the stage near the pulpit, called the entire assembly to attention. The general was dressed in full dress uniform with battle jacket. He had a silver star on each collar and each lapel.

Now he stood in front of the pulpit with his Bible under his right arm. His rugged face took on a serious expression, and his eyes flashed as he studied the faces of the men in front of him.

When the auditorium was in complete silence he commanded: "At ease and take your seats."

He carefully placed his Bible on the pulpit and opened it to a place marked with a white index card. He spoke in a loud voice that could be heard by all.

"My fellow prisoners. I have considered it a great privilege to be the leader of ten thousand of the most courageous soldiers ever to serve in the Armed Forces of the United States. I wish to congratulate you on a job well done and a job accomplished far beyond the call of duty. The United States has reason to be proud of each of you.

"On January 12th the Russian armies crossed the east German frontier, and they have been advancing steadily toward us ever since. I received a news report less than one hour ago placing the Russians less than twenty-two miles from this camp."

Noise and outcries filled the auditorium and the general sternly commanded:

"At ease. During this period of great trial as the Russians continued to advance, I gave you every news report. I expected you to conduct yourselves as soldiers and on several occasions I urged you to stop all demoralizing rumors. I pointed out that they were destroying morale and teamwork. I tried to show you that continuation of such idle talk would produce a weapon great enough to destroy us. My orders were repeatedly disobeyed. Now morale in this camp is at an all-time low. Your faith has been replaced by stupid fear. Your courage has been replaced by weakness. I cannot lead weak sisters. I am much older than most of you, and I want you to listen to me.

"For my short lifetime I have accomplished more than the average man. I am a general in the United States Army. I have been attached to the General Staff. I have served as assistant to the American ambassador in Berlin as military attaché to Berlin. I have dined with the leaders of Nazi Germany, and I watched them drive the British into the ocean at Dunkirk. I am supposed to be a leader and a diplomat. I was trained to understand the ways and methods of European people. I speak several foreign languages. I have also acquired substantial financial means and social position in the United States. All of this has taken me a lifetime of work and study."

He paused to allow his words to sink in. Then he continued:

"Now as I stand at the crossroads in life, all of this has been swept away. My friends, my money, my position, and my training cannot help me now. I am here with you to face my destiny. As I search desperately for help, I find that there is only one voice and one power that can help me; and that voice and power is Almighty God. If He refuses, I am lost.

"I am now going to tell you exactly what is going to happen to us. It will be one of three things, because that is all that can happen. One: The German guards will either evacuate or surrender this camp to the Russians. In that event we will be liberated with little worry. Two: The commandant will be ordered by some high fanatical official in Berlin to put us to death. In that event we must fight for our lives in hopes that some of us will be saved. Three: We will be evacuated on a long march across Germany. In that event we will suffer many casualties and it will be a March of Death.

"Our best chance for survival is to stand together as one team, ready to face whatever may come. God is our only hope and we must trust Him. We must also have confidence and complete faith. We must also be ready to do whatever is necessary if we want Him to help us.

"Now I want every man in this camp to return to his barracks, and at 11:00 a.m. we are going to hold one hour of prayer. You can read the Bible, pray on your bed, pray in groups, or pray aloud, but I want every man to pray. When this hour of prayer is over I shall expect you to be ready to face whatever may come."

Again he looked out over the audience and then down at his Bible. As his eyes swept over the auditorium he told us:

"I don't know about each of you, but for me I have this to say, in closing. 'Yea, though I, General Vandermann, walk through the valley of the shadow of death, *I* will fear no evil; for *Thou* art with me.'" He skipped a few verses of the Bible and concluded: "'Surely goodness and mercy shall follow me all the days of my life; and I shall dwell in the house of the Lord forever.'"

He closed his Bible and started toward the steps of the platform. Colonel Spivey called us to attention, and we stood in silence as the general departed through the main entrance.

At eleven o'clock all camp activities were closed down, and every Kriegie joined the members of his combine. Several thousand Bibles were opened and read. Some men read silently, some read aloud and others read to small groups. Some retired to their bunks and prayed silently in bed. Others knelt by their beds in prayer. A few men gathered at each combine table for group Bible readings and group prayers. Men who had never prayed aloud before led prayers that would never be forgotten. There was silence and seriousness everywhere.

When the hour of prayer came to an end, barracks doors were opened and we began to move about. A strange feeling of peace and security swept over the camp. Men were strong, full of laughter, and had absolute confidence. As I moved about the barracks, I heard men talking of the possible liberation or evacuation. There was no other consideration.

On Saturday, January 28, morale remained unusually high. Men walked the perimeter track and evacuation plans were resumed. The Russians were still reported to be approximately twenty-two miles from Sagan. In the early afternoon the rumble of artillery could be heard; we estimated that it was anywhere from fifteen to twenty miles away.

It was now the considered opinion of the large majority of us that evacuation was completely out of the question. It would take several hours to even march ten thousand prisoners out of Sagan, and at any moment the Russians could make a minor forward thrust and capture the camp. If the commandant tried to move us out, the Russians could easily encircle us and capture the German guards.

Saturday night most of us played poker, bridge and Monopoly waiting for the latest official report on the Russian locations. We were certain that twenty-four additional hours would end the war as far as we were concerned.

Around nine-thirty as we were finishing our coffee, Captain McGee stormed into the barracks. He stood on top of Combine D's table and shouted: "At ease. Gather around and let me get my breath." His face was chalk white, and his eyes were wild. No one had to tell us that something big had really happened, and that he was about to announce our fate. He lit a cigarette and started talking.

"I have just come from the auditorium. All block commanders were at the meeting with General Vandermann and the assistant commandant. We have been ordered to evacuate the camp immediately. We are to be ready to start marching in one hour."

Expressions of fear and panic ran about the room.

"Quiet," McGee shouted. "Please be quiet. I know you are just as shocked as I am, but we haven't a minute to lose. The commandant had intended to surrender the camp, but orders came from Berlin to evacuate Sagan immediately, and move the entire ten thousand prisoners in the direction of Berlin. Two officials of the S.S. are at the commandant's office now. The Russians are very near at hand, and the Germans are scared to death. They have been told that we are their only chance for survival, and that we must be held as hostages. They are extremely excited, and we have been advised to do nothing that might alarm them. This is a serious situation. We must carry out their orders to the letter. They have been ordered to shoot any man who hesitates or who fails to carry out orders. Roll your packs and prepare to depart as quickly and orderly as possible. Help everyone you can. Divide your combine food as we have practiced, and dress as heavily as you can. It is extremely cold, and it might start snowing before the night is over. It is now 9.40 p.m. We are to be ready to fall outside at 10.30 p.m. Every combine leader report to me as soon as your combine is ready to fall out. That's all."

The next forty minutes of accelerated activities surpassed all records for assembling and preparing for a mass-moving project. Closets were emptied, food divided, packs rolled, beds disassembled, kitchens stripped of food and dishes, tools and weapons were uncovered. Most of us dressed with two pairs of socks, two suits of underwear, two shirts, overcoat, hat and G.I. gloves. By ten-thirty most of us were ready and waiting for the order to fall out.

McGee called four men to his room, and I was one of them. "All right, men, we haven't a minute to lose. Each of you knows that you are to carry a major part of the camp record. Since you know it by heart, if there is the slightest sign that you are going to be searched during this march, destroy all documents. I suggest that you carry these documents inside your socks. They may search you without undressing you. Many times they will undress you, but allow you to leave your socks on. Do this now before you leave my room."

Each of us got our documents and removed one of our shoes and socks. We placed the documents inside and replaced the sock and shoe. As we returned to our respective combines, there was a message from the general.

"Gentlemen: Your departure has been delayed until 11:30 p.m. I am leaving you men of the Center Compound to lead this march. You will be the fourth compound in the order of departure, and that is the reason for your delay. Since you men of the Center Compound have maintained the Camp Record, my staff has selected certain men of the Center Compound to carry the record. Guard it with your life, and destroy it only in case of emergency. I will be with you until the end of the war. Carry out all German orders, and remember I will be looking out for your interest. Remember to keep your faith, your morale high, and don't forget your prayers. God bless all of you. General Vandermann."

As soon as this message was read, McGee came to the center of the block with another message from Colonel Spivey. He commanded "At ease" and read: "Men of the Center Compound: We will fall out at 11:30 p.m. prepared to march. I will lead the Center Compound from the gates of Sagan to the end of the march. Our destination at this time is unknown. Your guess is as good as mine. Certain men have been selected to carry the Camp Record. It is better if no one knows who these men are. If they get in a tight spot, they may call on you for help. Do everything possible to help them. The safety of this record is the responsibility of every man in this compound.

"I want to urge all of you to carry out orders promptly. Do not antagonize any German guard. They are heavily armed, and they have been ordered to shoot any man who breaks ranks, or who deliberately disobeys orders. They won't ask questions. Any man who attempts any kind of escape will be shot on the spot, and he will also endanger the lives of the rest of us. We will be safe, if we stick together as a unit. If the general deems it necessary, and if he is unable to negotiate with the commandant, he will give the order for a mass escape. In this event each man will be given an assignment. No one will act on his own. If escape is necessary, it will be done en masse. We would then probably take shelter and try to hold out until the Russians overtake us. Good luck, and remember to carry out orders promptly."

All of the men in my combine were in good spirits. We were

leaving a good many belongings behind, but we had the main essentials. We packed all of our food, clothing, cigarettes, and a few personal belongings and souvenirs. When Colonel Spivey's message was read, we still had almost thirty-five minutes to wait.

It was just past 11:30 p.m. when the goons entered the barracks. Eight privates and a sergeant of the guard were assigned to each block. They took up positions in the aisle, and we noticed they were heavily armed. Each private carried a pistol, a rifle fixed with bayonet, and two hand grenades.

The sergeant was carrying two pistols strapped to his waist, a cartridge belt, and a submachine gun. The sergeant was new to most of us, and we judged him to be in his middle fifties. He was short, stout, and wore a black mustache. He spoke excellent English. As the guards took up their positions in the barracks he ordered:

"Silence. I am Sergeant Von Ord. I have eight men with me, and I will be responsible for guarding you hundred men. I do not wish to see anyone get hurt, but it will be absolutely necessary for each of you to follow orders to the letter. Do not be foolish enough to even think that you have the slightest chance to overpower us. We are heavily armed, and we will fire on any man who makes the mistake of moving from the formation. We are going to march in a column of fives and guards will be on all sides of you. All extra sergeants not assigned to blocks are also armed with submachine guns. In case of a mass break, we have been ordered to fire on every man in the group. If one or two men break formation, the rest of you stay in line, and you won't get hurt. Everyone must keep up. We are going to march all night and all day tomorrow. The commandant says that if we are out of danger from the Russians, we will rest tomorrow night. If we are still in danger, we are to continue marching. We have been ordered to shoot any prisoner who falls out from exhaustion, and who cannot continue on the march. We cannot leave anyone behind. Again I want you to know that I am old, and I do not wish to hurt any of you men. I will not hurt anyone if you all obey my orders. Again, if you do not obey my orders, I will shoot you without hesitation."

We heard the bugle blow. Sergeant Von Ord gave the order to fall out and departed to the outside.

As we fell in block formation on the parade ground, we all

knew that this was our last appell at Sagan. In the center of the parade ground Von Ord and all other German block sergeants reported to Pop-Eye. Pop-Eye was the top sergeant of our compound, and he was to direct our march. We were all relieved to see Pop-Eye in charge. He was plenty tough at times, but we knew he was not fanatical and that he would not deliberately injure any of us. Pop-Eye issued the last-minute orders while the appell was being taken. Dumbkoff and Shisenkoff counted us correctly for the first time since I had come to Sagan.

Block one faced the north goal post with Colonel Spivey ten paces out in front. In a few minutes Pop-Eye joined him and gave the signal. Colonel Spivey gave the order and block one moved off. Ten minutes later we were marching through the front gate of the Center Compound in a column of fives, four hundred men deep.

I had not realized the bigness of Stalag Luft III until we marched clear across the camp. It seemed as if we had marched over half an hour and we were still in the North Compound. As we proceeded through the North Compound, we saw German guards pouring gasoline on roofs and sides of barracks. They were going to burn the camp, but we had expected that. Nothing was to be left for Russian troops to use as quarters.

We moved through the north gate of the North Compound down a small winding road toward the city of Sagan. It was very cold, and a light snow had started to fall.

I had never expected to be marching from Sagan in the direction of Berlin. I recalled that I had been captured just three and one-half months ago. It seemed like years ago, to me. Sagan had shown me a life of excitement and adventure. It had also been a life of development, growth, and change. In my food sack I was carrying the rolls of toilet paper on which I had written my diary. I had changed a great deal. All of the men marching with me had changed too. The mark made on us at Sagan would never be erased, and I knew it.

We moved up the long hillside. The column slowed down as the climb grew steeper. Somewhere near the top I turned around and saw Stalag Luft III in flames. The fire covered several square miles, and flames leaped high in the cold air. This was the end of Stalag Luft III and the beginning of the March of Death in Germany.

XIX

The March of Death

THE NIGHT WAS COAL BLACK, THE MOON HAD DISAPPEARED, AND millions of tiny snowflakes slowly drifted toward the ground. The snow was stacked two feet deep, and more of it continued to fall all over our part of northern Germany. Along, endless column of black figures moved slowly forward through the crowded streets of Sagan. As the column continued to move, German guards ran their flashlights over the faces of American prisoners, checking the formation.

German civilians cleared the center of the road as the formation passed by. Thousands of civilians gathered along the road and moved slowly out of Sagan toward Berlin. Some were in wagons, but the large majority were walking. Each civilian carried several bundles; many had more than they could handle.

We watched in silence as soldiers of the German army and S.S. hurried the civilians into the endless line of marchers. The S.S. troops were setting fire to business buildings and homes and directing civilian traffic. Everyone was to be evacuated, and nothing was to be left behind for the Russians. German civilians who resisted were shot on the spot and left in the snow; we heard the screams and shots. The S.S. never argued; a rifle shot saved time and settled all arguments.

By two-thirty in the morning our column was moving along the main road two miles north of the city of Sagan. As prisoners, we were in far better shape than the civilians about us. The sight was pitiful and depressing. Civilians marched on each side of us on the gravel shoulders of the highway. Old men were being helped by young boys. Mothers were carrying babies, and young children were carrying bundles of belongings and food. Many were poorly clothed and half frozen from the bitter cold. Babies wailed and mothers wept as we plodded slowly forward.

Our travel along the highway was almost paralyzed by the thousands of people moving in one direction. These civilians were bitter and demoralized, and many of them, I felt sure, would rather surrender to the Russians than continue. The civilians were, in fact, prisoners too, for the S.S. troops moved them by the force of rifles and bayonets. Every now and then some civilian would sit down and refuse to go on. A rifle shot allowed the civilian to sit there forever.

We kept close formation and maintained the best possible morale. With each forward mile, more and more civilians joined the endless line of marchers. German morale was near the breaking-point, and we knew that the end was near for Hitler's Germany.

The snow continued to fall, and by five o'clock thousands of us had numb faces. Our packs were growing heavy, our bodies were getting weary, our feet were tired from plodding through the snow, and each mile of advance became slower and more difficult.

Shortly after eight a.m. the Germans ordered a fifteen-minute rest. We were allowed to rest on one side of the road while the goon guards watched us from the other side. We dumped our packs and flopped in the snow. Food packs were hurriedly opened, and members of combines gathered in circles to eat their breakfast. Eating in the snow was cold, but after eight hours of continuous marching none of us seemed to care. We were too hungry, tired, and cold to worry about getting colder. It was during this breakfast that I realized this march would indeed be a March of Death and the greatest ordeal of my life. While we were still eating, the whistle blew to fall in again. We gathered our belongings, fell in formation, and moved off, eating as we marched.

The snowstorm continued for hours, and the formation moved slower and slower. We were drenched to the skin. It seemed at times that we were stationary. The colonel had every block commander talking and trying to cheer us up. Songs were started but never finished. We were too cold to open our mouths. Humor ceased. Bitterness mounted. By noon the thickness of the falling snow made it difficult to see beyond the man in front of you.

In the early afternoon several German guards, who were middle-aged, collapsed in the snow. Pop-Eye had them taken to the rear and loaded on the German bread wagons. We continued marching all day long, stopping twice for a five-minute break—just time enough to smoke a cigarette. Everybody was cursing, including the German guards. To all of us the commandant was public enemy number one.

Just past four p.m. we entered the small town of Wharton and stopped for a break. I had never been more exhausted in my life, and I felt sure that one more mile would put me out of action. Many Kriegies were really getting sick.

During this break, Colonel Spivey announced that General Vandermann had sent the British Compound on toward Muskau. He told us the general was having a dispute with the commandant about going further in the storm. Thirty minutes later Colonel Spivey returned and told us that General Vandermann had refused to go further without an overnight stop. He and the commandant had had a furious argument. Each had threatened the other. General Vandermann had stood firm, and the commandant finally had ordered an overnight stop.

The Center Compound was assigned to a Roman Catholic church capable of seating four hundred people. It took one hour and forty-five minutes to pack the two thousand members of the Center Compound into the small church interior. On the main floor, Colonel Spivey's staff packed eight hundred fifty men in the pews. Men were jammed against each other without room to lift or move their arms. After the pews were filled, every fourth man moved under the pews and spread his blanket to stretch out for a rest. This schedule under the pews was to be rotated every three hours.

The pulpit and choir section was filled, and three men were assigned to each step leading to the choir. Directly in front of the pulpit, twenty-three men were assigned to the kneeling rail. There were four aisles on the main floor besides a large space between the pulpit and the first rows of pews. Men were lined up and seated on each side of the aisles from the front to the back of the church. The center of each aisle was left open for those who had to move about. Forty-eight men were seated on each side of the four aisles. The space in front of the pulpit was used as a sick-bay for the sickest men who were in need of constant medical attention. One hundred and eleven men were assigned to the small basement and storeroom. All washrooms, toilets, and exits were filled to capacity.

Three hundred men were jammed in the balcony. The same procedure was used in filling balcony pews and aisles. There were two sets of stairs leading to the balcony, and two men were assigned to each step. Seventy-five of the sickest men were finally able to lie down when all men of the Center Compound had been assigned space.

When our shelter arrangements had been completed, members of Colonel Spivey's staff were issued first-aid and medical kits to attend the sick. A sick-bay was formed in the home of the Catholic priest, which adjoined the church, but it was abandoned when five hundred men of the Center Compound reported for medical treatment. The most severe cases remained in the home of the priest, and all other sick men were returned to the church. Approximately two hundred men were suffering from frostbite. Over one hundred men were in varying degrees of shock. Three dozen Kriegies were diagnosed as pneumonia cases. Several hundred others had severe cases of dysentery and diarrhea.

Our situation was much worse than we had expected. With one day's marching completed, thirty per cent of the men of the Center Compound were sick in varying degrees. We all realized that several more days of similar marching conditions would make it impossible for us to continue. Our medical supplies were totally used up within two hours after the sick-bay was established in the church. It was bad enough not having doctors to attend the sick, but from now on we would be without medicine too.

The heat from two thousand bodies warmed the small church. Most of us removed our heavy clothing, hoping it would dry before the march was resumed. Several large tubs of snow were brought into the church. They were to be used to refill canteens and water jars when the snow melted.

Several men from each combine unrolled the food packs and prepared rations for the combine to eat. One man from each combine got the water ration. Those men who were not fixing food, dried clothes and helped the sick.

Rhinehart and Mott fixed twelve spam sandwiches. They passed each of us a sandwich plus a piece of cheese and a small block of chocolate. Tom Pirtle filled each tin cup with cold water. Our combine was housed in the balcony. I looked out over the balcony and saw hundreds of men sitting, standing, and crawling about. Food was being passed up and down the line. After the meal, we started talking for the first time in six hours.

"How in the hell Colonel Spivey ever got two thousand of us into this small church is beyond me," Ted Carson said.

"Thank God, we are inside. That is the worst snowstorm I have ever seen. How we made it this far is more than I can understand," Rhinehart continued.

"I agree with you, Bob," Pirtle said. "Those damn Germans are having it plenty rough themselves. If this weather keeps up, this march isn't going to last much longer. A few more days of this and none of us will be able to walk."

Dave Roberts stretched against the balcony railing and said: "We better try and get some sleep. If we have to march tomorrow, we are going to need it." We finally settled down for the night.

The church was packed so that any man who found it necessary to move more than a few feet was almost certain to step on someone else. The snow tubs were placed in the exits, and men were constantly moving to the exits for water. Toilets were set up outside, and there was a continuous line to and from them. With two thousand men in one small building, lines were moving all night long. Many men became desperately sick at their stomachs and were never able to reach the door. Dozens of men rushed up aisles vomiting all the way. Others with dysentery stepped on hands, feet, and stomachs, trying to get outside. Nerves were strained to the breaking point.

By ten o'clock men were cursing, crying, and fighting. Disorder sprang up over the church, and no one could sleep for fear of being puked on. The odor in the church had grown strong enough to make all of us sick.

We heard someone shouting "At ease," and finally the church got quiet. Colonel Spivey and Chaplain Daniel were both on the stage in front of the pulpit, in their underwear.

"When I call 'At ease,' I want every man to shut his damn mouth," the colonel said. "If I catch another man fighting, he will stand outside in the snow the rest of the night. This situation has created a real crisis and you are not acting like men who are capable of facing a desperate situation. Let me tell you now that being pushed, shoved, stepped on, even vomited on is far from the worst that could happen. We are inside now, and less than three hours ago we were outside freezing to death. We are warm, and we have been given water and food. It is going to get a lot rougher than this, so you might as well prepare yourselves. Any man who could strike a Kriegie who is sick is nothing short of a coward, and I mean that. If you are going to start fighting among yourselves tonight, then I ask you, what will tomorrow bring? I'll tell you what it will bring. It will bring more fights, and finally a revolt among ourselves. That is just what the Germans want. A revolt is justifiable reason to shoot every one of us. The German guards feel that we are excess baggage, and they are looking for an excuse to get rid of as many of us as they can. I have arranged you in this church to the best of my ability. The sickest men are nearest the doors. You must remember that you may be sick yourself before this march is over.

"Now, I don't know about the rest of you, but I am grateful for this shelter, vomit, puke, and all. Any man who feels he just can't stand any more will dress and fall outside. I want you to stand out there in the snow at two degrees below zero. Keep telling yourself that you won't come back in unless you can stand the vomit and other discomforts. I'll bet a carton of cigarettes that not one of you will stay outside for over thirty minutes. Anybody want to bet? All right. I want all of you to try to sleep. I ask everyone to make as few trips as possible. If you see that a man is sick, help him get outside. Be courteous to your neighbor. If you can't sleep, sit there and dream of home.

If you can't say something pleasant, keep your damn mouth shut. Good night."

Colonel Spivey left the stage as Chaplain Daniel stood in front of the pulpit. He smiled across the audience, and many men smiled back at him. He spoke in a soft voice.

"Surely we have not forgotten that today is Sunday, January 29, 1945. God sometimes acts in strange and mysterious ways. He did not see fit for us to be liberated at Sagan. We have been through a great hardship today, but still we have been cared for. We are alive, we have eaten, and we have shelter. Did you ever stop to think that maybe God is testing our faith? God is here tonight, and He is watching us. Let's show Him that we can endure whatever may come our way. Everyone bow his head while I pray. Almighty God, we thank Thee for our lives, for the food we have eaten, and for the shelter this night. Bless and protect those among us who are sick and weary. Cure them, oh Lord, for we are out of medicine. You are the only doctor we have left. Help us tomorrow as we march onward. Give us the necessary strength to survive and go ever forward until we find freedom and liberation. Amen."

Chaplain Daniel was to us someone special like General Vandermann. He was a Protestant minister, a handsome man no more than twenty-five years old—one of the men I shall never forget. He always had a smile; his good nature, cheerfulness, and tolerance were unfailing. And what understanding this young, quiet, and even reserved man had.

How much we all owed to him on this March of Death cannot be expressed. He was constantly dropping back along the columns of prisoners to visit, talk to, and encourage the struggling men. During the rest periods he roamed through the groups attending to the sick and praying. It was amazing where he got the energy to perform his good works. He was warm and able to reach the hearts of his fellow prisoners.

What Chaplain Daniel had, and it must have been his source of strength, was an unlimited faith in God which he conveyed to us. He never had the slightest doubt that we would survive the tremendous ordeal of our march, and he made us feel that we would come through.

We finally settled down for the night. I must have slept for several hours. I woke up wringing wet with sweat and deathly sick at my stomach. I moved carefully down the steps to the front door. I puked until there was nothing left in me. I went to the toilet three times during the next hour, but the pains in my stomach steadily increased. When I finally got back to my seat, Byron Clark woke up and assisted me.

"I am sick," I told him. "The pains in my stomach are killing me."

"Here, try to sip some water." Clark felt my head and told me I had a high fever. I was burning up. I drank several cans of water trying to put out the fire in my stomach. Thirty minutes later I got sick again, and barely made it to the front door. Clark had to help me back to the balcony. My fever was mounting and an hour later I was out of my head. I saw blurred images about me but I couldn't recognize anyone. Someone wrapped me in a blanket and sweat poured from my body.

When I woke up at nine the next morning, I was downstairs lying in the sick-bay in front of the pulpit. I tried to sit up, but couldn't. I was too dizzy to recognize anyone. Someone was holding a pan for me to spit in, and I noticed blood. If I were spitting blood, I must be worse off than I realized. For several minutes I would burn inside and perspiration would cover my body. They would remove the blanket; in a few minutes my teeth would chatter, and I would start having a chill. I had lost track of time, but I knew the men were falling out outside.

Promptly at ten the German guards took up their positions outside the iron gates in front of the church, and the bugle sounded formation. Men came from within the church carrying packs and helping sick comrades take their places in the column.

Clark and Coletti helped me to my feet. I stood up with their support and looked about. My head was swimming.

"Ken, you are sick," Clark said, "but you must fall out with us. The goons will shoot anyone who tries to stay behind. We are going to get you in the column, and if you can't walk by yourself all of us will help you. Gene and Tom will be on each side of you, and if you think you are going to fall, hold on to them. Do you understand what I am saying?"

"Yes," I answered, "but I don't think I can make it. I am so weak I don't believe I can walk ten feet."

"Oh, yes, you can," Gene said. "You can make it all the way.

Just remember, Ken, we are on our way home. Just a few more days at the most, and you will be on your way home. Keep thinking of Texas, and you will make it."

I stumbled several times before we reached the formation. I knew I must have pneumonia, for my lungs ached, but I was determined to keep going as long as possible. When I got in the formation I started shaking all over.

By the time the compound was formed, forty-nine men were still inside the church too sick to be moved.

The commandant approached Colonel Spivey and said: "Colonel, I will wait just ten more minutes for you to get those men outside. After that we are leaving."

"Listen, Major," Spivey shouted. "Don't you know that those men are too sick to be moved? Couldn't you put them in a wagon or truck? They can't possibly march in this weather. They will be dead before noon."

"Look, Colonel Spivey, we have our own problems. I have over twenty sick guards, and the wagons are loaded. Those men cannot be left behind. They will either fall out ready to march, or they will be left behind to be shot. Those are my orders from the S.S., and there is nothing I can do about it."

"If any of those men are shot, Major, I promise that I will watch you hang from the highest gallows," the colonel said.

"Oust," the major shouted. "They either fall out, or they will be shot."

Colonel Spivey stormed to the main entrance of the church with two men from each combine of the men who were still inside. Captain Randolph was with the colonel, and he did the talking. When they were inside, the door was closed. Randolph stood in front of the thirty-nine men lying on the floor and said:

"You men lying on the floor, sit up and listen to me."

All of the men opened their eyes. Some sat up and others propped themselves on their elbows.

"You men are sick, but you are not near as sick as you think you are. If you think you are in bad shape, then you ought to see those German civilians out there on the highway. We have exactly ten minutes to get you outside. Every one of you is alive, and we have men here from your combine to help you get in formation and start marching. Maybe you think you can't make it, and maybe you can't; but if you stay here on this floor your worries are over, because the Germans are going to shoot you. I have seen men a lot sicker than any of you, endure a lot more than you are being asked to endure. Half the battle of living is wanting to live. If you really want to live, you can do the impossible. You had better start thinking it over and make up your minds fast. You are just a few days away from freedom and the end of the war. The Russians are bound to overtake us at any moment, and then we will start our journey home.

"Do you hear me?" he shouted. "We are going home. Are you going to lie there on that floor when the rest of us are going home? Think of the United States. You can't forget a life like that. Think of your family, your friends. You will have a good place to live. You can have a complete new wardrobe of clothes. You can drive a car again. Think of all of the beautiful women back home. Women, I said."

Several men started rising to their feet. The members of their combines gathered their belongings and helped them to the door. As they started getting up, others helped them the rest of the way. Captain Randolph steadily raised his voice, almost shouting.

"Think of all that money. What are you going to do with all that money? Most of you have over five thousand dollars saved. Are you going to let somebody else spend that hard-earned money? If you stay on that floor you are going to die and somebody else will have a big time with your money." Those lying on the floor now really came to life. Over a dozen got up when he talked about money.

"Freedom is in your very grasp," he went on. "Freedom forever. Freedom with plenty of women and money and the best place in the world to spend it. I really thought all of you had guts. If you really want to go home, get up off that floor. Your combine members will help you get started. If you go out that door and fall dead before we move off, you are better off than if you give up and lie here to be shot. Now when you get in the formation, I want you to keep repeating this over and over. Every time you take a step I want you to say: 'I'm one step nearer home, and I am going to take another. I am one step nearer home, and I am going to take another.' Keep saying that as we march, and I know you will make it. All right, help those nine remaining men to their feet. We have got two minutes before time to move off."

Colonel Spivey watched in amazement as they staggered, fell, and stumbled to the door. Two Kriegies helped each of them as they passed outside.

When the formation was ready to move off, a dozen men were being helped to their feet. Men on either side of them wrapped their arms around their necks and helped them stumble forward.

I started marching with my head down so that I could see where to put my feet. I heard a man several places behind me say: "I am one step nearer home, and I am going to take another."

After an hour's marching, I was walking on my own with my head up. I must have said it a thousand times, "I am one step nearer home, and I am going to take another."

We stopped on the side of the highway at three p.m. for a lunch break. I had been too sick to notice the thousands of civilians marching with us. I looked forward of the column toward the long sloping hills several miles ahead. A tiny black column was moving forward as far as I could see. I looked to the rear, and saw an endless line of human beings moving in our direction. This was January 30, 1945.

Rhinehart fixed our lunch, but I was too sick to eat. I sat by Gene and Tom Pirtle and ate some snow. It was refreshing and cooled my burning stomach.

Just after the whistle blew to fall in, one of the men from block eight, who had been helped from the church, fell face down in the snow. Captain Randolph rushed to him. In less than a minute Colonel Spivey and Chaplain Daniel were by his side. He was dead when they turned him over.

His body was carried to the wagons in the rear of the column. They would bury him when we made our next overnight stop.

The march was resumed, and during the afternoon I became dizzy several times. Once I nearly fainted. I kept praying and saying, "I am one step nearer home, and I am going to take another." Sheer determination kept me going. We continued to move in the direction of Muskau.

Before we got another mile down the highway it started snowing again. It came down slowly at first, increasing steadily until the sky, air, and ground were heavy with snow. The wind started blowing, and snow was everywhere. The long columns began to slow down, and marching got slower and slower. With all the snow, there was still that endless talking and squealing in German. Babies cried, mothers moaned, and old men cursed and grumbled. The civilians cursed Hitler, the army, the S.S., the American Luft gangsters, and themselves. They would fall, stumble, and even sit down, but when an S.S. trooper would appear with a bayonet at the end of his rifle, they were always ready to start marching again.

I decided the Russians would never overtake us. I wondered what could have happened to them. I knew the snowstorms were slowing them down, but surely they could move several times faster than we were moving.

During the afternoon, we marched six miles without stopping. The storm was at the height of its fury, and progress was extremely slow. Colonel Spivey kept marching up and down the columns giving words of encouragement.

"Keep going, men. We are going to stop again overnight. It won't be much longer. Just a few more miles to go. Keep your ranks closed in. If any man falls, help him the rest of the way. It won't last much longer. Think of your money. Keep going, and you will get to spend it."

How he managed to act as if he were on an overnight hike was more than I could understand.

Finally at 5:30 p.m. we stopped for a break. Word was passed through the ranks: we would stop overnight at a small German farm village three miles up the road.

When the march was again resumed I was so exhausted I didn't know what I was doing. I stumbled onward and kept repeating: "I am one step nearer home, and I am going to take another." I had to keep going. Everybody seemed to be counting on me.

Gene knew I was about finished by the exhausting snowstorm and he kept talking me on.

"Come on, Ken. I know you can do it. Just one more mile to go. Think of it. You have walked all day long, and you are still walking. One more mile, and then you can sleep all night. I am going to pour hot water down you all night, and I know you will be much better in the morning. Pirtle said he was going to give you some real hot coffee when we get inside."

"Just one step nearer home, and I am going to take another." I was repeating it over and over when the formation came to a halt. I looked up. It was dark. There were several large barns and farmhouses about us.

German guards were carrying lanterns and directing the assignment of quarters. My block and block eight were housed in one of the large barns. Outside every barn several German women had large fires going, and boiling pots were being heated with water. Many of them wanted to trade potatoes, onions, and bread for cigarettes, chocolate, and soap. The commandant made arrangements with Colonel Spivey whereby one member from each combine could come out and trade with the German housewives.

Our barn was approximately twice the size of the Catholic church. With block eight and our block to share the entire barn, we had room to spare. Hay was piled high on each side of the barn and it was warm inside. Each combine was assigned a large section of the hay, and the men of each combine gathered their belongings and food to prepare their quarters and meal.

Blankets were unrolled, and beds were prepared for the night. Many men decided to sleep in the hay. The hay was fifteen feet deep on both sides of the barn and no one had to worry about reaching bottom. Each man who slept in the hay dug out his bed and packed it with his feet and hands.

Combine C gathered in the far north corner of the barn. I stretched out on the hay, too sick to do anything. Clark got all twelve members located, and each man got his bed ready.

Bob Rhinehart and Bill Mott couldn't wait to start trading with the German frauleins. They drew straws, and Mott won. He took three bars of soap, five packs of cigarettes, and one bar of chocolate with him. In fifteen minutes he returned, smiling and singing.

"I've made the deal of the century. Pirtle, get the coffee, and Rhinehart, give me some of that spam. I traded two old women a package of cigarettes for a ten-gallon pot of boiling water. When they saw the soap, they went wild. You know it surprises me how well they speak English. They have gone after some potatoes, onions, and meat. They also said they would spice it."

"Spice what?" Pirtle asked.

"The stew! They are going to bring a smaller pot back and make us a three-gallon stew. The rest of the water will be for coffee and whatever we need."

"How much did all of this cost us?" Pirtle asked.

"Three bars of soap and three more packs of cigarettes. Here is the chocolate and the other pack of cigarettes."

Rhinehart passed ten potatoes and a full can of spam to Mott. Pirtle gave him the coffee jar. Bill loaded his pockets and started across the hay to the floor.

For the next thirty minutes all of us undressed and hung our clothes on rafters and nails to dry. We stripped naked, and after all clothes were hung out we put on our overcoats and buried our feet in the hay.

I still had cramping pains in my stomach, and dizzy spells hit me at regular intervals. I was beginning to get warm for the first time since we started the day's march.

When Mott returned with the large pot of coffee, the combine cheered him.

He laughed and told us: "All right, you bastards. Fill your puny guts with coffee. I am saving my space for stew."

He started to leave when Pirtle shouted: "Did those women come back?"

"Of course," Bill answered. "They have got more stuff in that stew than you ever dreamed of. They are trying to talk me out of my own cigarettes. Another thirty minutes ought to have her ready to serve."

Gene got my tin cup, and fixed me a boiling cup of coffee. I sat up in the hay sipping it. It burned all the way to the bottom of my stomach. While we were waiting for the stew, I drank a can and a half of coffee.

Mott came back after the coffee pot, and returned with a coffee pot full of stew. Rhinehart opened the lid of the pot and dipped it out with a cup. He filled each tin cup brimming full. We spooned it, blew, sipped, and ate it as it began to cool. I had never tasted anything better in my life. It was well seasoned and very thick. When I had consumed two cans, my stomach was stretching and I was sweating all over.

I buried myself in the hay and used my overcoat as a blanket. I lay there for an hour listening to the conversation and sweating. I was wet from head to foot. Buried in that hay was like a Turkish bath. I finally went to sleep talking to Gene.

About three o'clock I woke up, still wet with perspiration. I got my overcoat on and made my way to the floor and exit. The toilet had been established by the north end of the barn. I was still having dizzy spells, but they were less frequent. My lungs still hurt, and at times it pained me to breathe. I hadn't smoked a cigarette all day, and when I got back inside I smoked one by the door. I knew I shouldn't be smoking, but I felt it might help settle my nerves.

When I got back into bed, I prayed for a long time. When I went to sleep, I was still praying.

I woke up at nine o'clock, and most of the men were still sunk in the hay. Ted Carson told me that Mott had saved about a gallon of the stew, and was outside warming it up for breakfast. Ted brought my clothes, and we dressed together.

"Ken, how are you feeling?" Ted asked.

"I feel much better. I am still weak, but my lungs have quit hurting. Ted, I woke up last night, and before I got back to bed I thought I was going to die. I had a feeling that I was at the end of the road. I must have prayed for at least an hour. I can't exactly explain it, but the pain gradually left me as I prayed, and I went to sleep praying. This morning when I woke up I took one deep breath, and I knew my prayer had been answered. A man just doesn't recover from pneumonia in my condition, sleeping in a pile of hay."

"No, Ken. It is against the laws of Nature. Yesterday, I watched three of those men the colonel brought from the church. They were walking, but they looked exactly as if they were dead. It scared me to death just to look at them. Every one of them was at death's door. Then they were brought out into zero weather in the snow. I wouldn't have believed it, if I hadn't seen it with my own eyes, that any one of the three of them could have walked one single mile. They stumbled and fell, barely holding on to the last breath of life. They kept going, and only God knows how they did it. Chaplain Daniel was walking by our combine when the snowstorm hit. He kept watching them and praying. He prayed and prayed, and before we took our last break I heard him say, 'Save them, oh God. They have had the courage of Thy Son.' They marched right through that entire storm and right to this barn. Look over there at Combine H of block eight across from us. Those three men dressing in the corner are the three men I have been talking about. When I saw them a while ago I nearly collapsed from shock. Those three men were as good as dead, and now they are standing there as if it never happened. Of course, God swept away the laws of Nature and saved them. What else could have saved them?"

While we were talking and dressing Captain Randolph came into the barn and barked out: "At ease. I have a special announcement to make. General Vandermann, the commandant, and the four American Compound Commanders have been in a meeting up at the village for two hours. The commandant has sixty-three guards too sick to be moved. Besides this, five guards have deserted. The commandant is also sick. He has agreed to another day's stay-over in the barn. It is still snowing outside. At three-thirty this afternoon Chaplain Daniel will hold a special prayer service for the Center Compound in the barn next to you. I think every man should attend. I also want to suggest that you sleep, eat and rest as much as you can. When we leave here tomorrow we are going all the way to Muskau. The general agreed to march there tomorrow, so you are going to need the rest. That's it."

Clark came over, and asked us to join him for a morning walk.

"Well, Ken, you look a thousand percent better," he said.

"Yes, and I feel a thousand percent better."

We walked around the barn several times, and talked about the march and the war. We all agreed that neither could last much longer. I began to feel weak again and returned to my bed of hay. The dysentery started working again and I made two trips to the toilet. Gene joined me for a talk and I thanked him for his help and service while I was sick.

"Ken, you are going to be o.k., now. This march has changed my mind about lots of things. For some strange reason I actually believe we are going to survive. If we can come through two days like the last two, we can make anything."

We talked for over an hour and then took a nap.

Clark woke us up for the prayer service, and the entire combine went as a group.

Instead of holding the prayer service in the barn next to us, it was held outside between the two barns. Chaplain Daniel stood on a small platform of four wooden boxes nailed together. I have never seen a man with such an expression. His face was all aglow. He was smiling and extremely happy.

"We have gathered here to thank God for our lives. For the past two days we have marched as prisoners of war across northern Germany in a severe snowstorm. We were weak when we started, and we are certainly much weaker now. We have eaten only four meals, we have done without sleep, and we have marched many miles under the most adverse conditions. Yesterday morning in this compound five hundred men were sick. Some men were critically ill and not expected to live. We marched in a snowstorm in sub-zero weather. We kept our faith at all costs, and God remembered us. He answered our prayers and delivered us to these barns and this village. He performed many miracles as all of you have witnessed, and he saved the lives of many. Whatever may come in the future, we are prepared to face it. After what you saw yesterday, surely no man will dare to doubt the power and willingness of God to help us. We give thanks to God for all of the blessings he has shed down upon us. We ask him to guide us through the darkness in order that we may find freedom. I want all of you to join with me in this prayer, and then I want you to remain kneeling and pray silently for five minutes."

As Chaplain Daniel began, hundreds joined in: "Our Father, Which art in heaven, hallowed be Thy name . . ."

A silence swept over the snow-covered ground, as men prayed. The snow had stopped falling, and the sun had come out. The warmth of the sun beamed down and warmed us for the first time in days.

When Chaplain Daniel rose to his feet he faced the sun. "Today is the last day of January. Tomorrow starts a new month and it holds new hope. Let us depart from this service by singing *Onward Christian Soldiers*." Two thousand male voices echoed through the air. Several German farmers came out of their barns and listened. It was truly a sight to behold. Two thousand American prisoners of war, on a death march in Germany, singing a great hymn to God.

XX

The Halt at Muskau

DURING THE REST OF THE AFTERNOON (JANUARY 31), COMBINE cooks were busy trading cigarettes, and soap, for hot water and food. Every combine tried to have its meal prepared by dark. Some of the German frauleins told the cooks that they hadn't smoked a cigarette in months. A package of American cigarettes was a rare possession among German civilians.

Immediately after the prayer service, I started having chills and running fever. Gene helped me undress and covered me in hay. Then he brought me cans of boiling water. The can was so hot he wrapped it in a cloth. I drank the water as hot as I could stand to swallow it; as soon as I finished one can, he would bring me another. The sweat continued to pour from my body, and I had to go to the toilet every thirty minutes. He got me a large jar for a urinal so I could stay in bed and keep warm under the hay. After I had eaten a can of stew I felt exhausted. I went to sleep early.

Captain McGee woke us at six A.M. We were to eat, pack, and be ready to march at seven.

The hot liquids worked like medicine on my body. My fever went, and I felt almost normal. True, my stomach was sore, but there were no pains. After I had dressed, packed, and consumed two hot cups of coffee, I felt like a new man. Still weak and shaky, I knew I was over the main sickness.

McGee came to the center of the barn and commanded: "At ease! You will fall out in front of the barn prepared to move out. It has stopped snowing, and the sun is coming up. We are going to march to Muskau, which is twenty-nine kilometers from here. The commandant is riding ahead in a staff car to arrange for our billets. We will spend the night at Muskau and from there we will go to Spremberg. A general in the Luftwaffe will meet the commandant at Muskau to give him final orders as to our destination. Since it is not snowing, we want to march as fast as possible. When we get to Muskau the march will be over for the day, so let's get there as fast as we can. All right, fall out."

When we were in block formation, Dumbkoff and Shisenkoff started taking the count. We hadn't seen either of them for two days, but they seemed their usual selves. Pop-Eye informed us that sixteen guards had to be left behind, and that he didn't want any monkey business.

During our stay in the farm village, dozens of Kriegies had traded cigarettes, soap, and chocolate for all types of sleds. Nearly every sled in the village had been sold to us, and dozens of new sleds had even been made by the German farmers while we were there. The sleds were purchased by an entire combine, because one sled was large enough to carry all packs, including the combine food packs. Clark bought our sled for six packs of cigarettes and three bars of soap. The sleds were easily pulled by two men. Our combine had six sled teams. Each team pulled for one hour and walked in the formation for five hours. The sleds made marching much easier, and we were relieved from the constant strain of a back pack. The sleds lengthened the compound formation, but Pop-Eye was ready for any deal that would be of help to him. Therefore we carried the packs of all the German guards. This was a great relief to the guards, since they were also carrying pistols, rifles, submachine guns, and ammunition.

At a highway intersection on the outskirts of the farm village we were halted until the West Compound took up their original position in the march. German civilians by the thousands were marching beside us. The civilians had also gotten wise, and most of them were pulling sleds and wagons. The sleds were loaded with clothes and goods, and many of them had small children strapped on top of the clothing bundles.

We were not allowed to talk to the civilians, but we nevertheless learned that many of the troops in charge of their evacuation had deserted. We also learned that large groups of civilians who were housed in barns overnight were not to be found the following morning. These were positive indications that many Germans were determined to remain behind and surrender to the Russians rather than to continue the futile struggle. Many S.S. troops ignored their orders and refused to shoot civilians who stayed behind.

We moved along the highway to Muskau in good marching order. The sky was clear, and the sun beamed down on us.

I improved steadily with each continued mile of marching. After we had marched several miles, the soreness in my limbs began to disappear. I thanked God, for I knew I had recovered.

Before noon, word was passed down the line that Wharton was to be evacuated within twenty-four hours. We knew that the Russians were moving once again, because hundreds upon hundreds of German aircraft passed overhead. All day long German planes streamed across the sky, all moving in our direction. To men in the air force, this could mean but one thing. Many German airfields were being evacuated, and the planes were being relocated at bases out of immediate danger from the Russian advance. We identified German trainers, transports, fighters, bombers, and observation planes as the retreat of German aircraft continued across the sky. More planes flew over us that day than we had thought still remained in all of Germany.

Every indication pointed to German defeat and surrender. The Russians were still advancing across Germany. Everyone knew the answer.

Peter Glass attracted a great deal of attention from the civilians. Many of them had never seen a Negro, and some seemed very interested in studying his appearance. One young fraulein pointed at Peter and shouted: "Colored American." Those that hadn't noticed him got a good look. It didn't bother Peter. He smiled and laughed all the time. Peter was walking on my right and Ted Carson on my left.

"Hell, Peter," Ted said, "they are too ignorant to know what it's all about."

"You can't blame them, Ted," Peter answered. "They have never seen anyone like me, so naturally they are curious. I am something new to them, and they want to get a good look."

A formation of bombers could be heard in the distance behind us. They were flying low at about four thousand feet altitude, and were moving fast. Pop-Eye stared hard for about five seconds, and then started screaming, "Clear the highway and take to the ditches."

He wasn't the only German who recognized Russian bombers, because German civilians were scattering in every direction before he opened his mouth. We didn't look back or wait for Pop-Eye to tell us a second time. We broke ranks and left the highway from both sides. It was a rat race as thousands of us spread across the fields. Every man for himself. People were knocked down coming and going. Everybody looked desperately for some place to hide, but open fields covered deep in snow offered the only available shelter. Twenty miles of highway lanes covered with human beings were deserted in seconds. I never saw fifty thousand people scatter faster in my life. The German guards forgot about their duty and ran like hell. Many of them passed us racing through the fields.

People hugged the ground all around us as the formation of fifty Russian bombers passed overhead. They were really flying low, but they were after something more valuable, for they continued in a northerly direction. As soon as they were safely in front of us Pop-Eye was back on the highway shouting orders. He could see for miles around him, and he knew none of us would think of escaping in this area; however, he regrouped his guards and took an appell. Dumbkoff was griping and complaining as we moved off.

We stopped around noon for a fifteen-minute lunch break. It was just time enough to fix a cheese sandwich, drink a can of water, and smoke a cigarette.

During each break it was necessary for several thousand Kriegies and civilians to use the toilet. If there was a wooded area near at hand, German women would occupy the woods; and the men would use the shoulders of the highway. Sometimes, however, there wasn't a tree in sight. In this event we Kriegies would use one side of the highway with our backs toward the civilians. German men would form a line on the shoulder of the highway facing us. German women would get behind the men and relieve themselves. Shy people tried to wait for stops near wooded areas, but sometimes the laws of Nature were too much for them. Even on a march of death most people were cooperative and considerate. Most German women tried to be liberal in these matters, but sometimes they would catch a peeker. The women did a good job of cursing peekers.

Shortly after lunch, we crossed several bridges which passed over small streams of water. Dumbkoff had dropped to the rear of the formation, and as we passed over the second bridge he threw his rifle over the side. He intended to tell Pop-Eye that he had left it at the farming village. He should have been more careful, for Pop-Eye got a glimpse of the rifle in the air. For a man with one eye Pop-Eye's observations were amazing. He beat Dumbkoff over the head with a club and put him out of commission for the rest of the day. Dumbkoff had a bump on his head, but he got to ride on the bread wagons the rest of the day. Bill Mott was marching in the column in front of us and Dumbkoff's mishap gave him his first real charge since the march had begun.

"Dear old Dumbkoff," he said. "He just didn't want to carry that big old heavy rifle any more. If he had kept his eyes on Pop-Eye he would have got by with it. I didn't know he had any guts. At least, he is smarter than most of the goons. He is back there sitting on our bread, while the rest of them are out here marching with us. I wish to recommend that we consider changing his name from Dumbkoff to Smartkoff. He has displayed real courage in the face of the enemy, and I shall request a combine vote after you have had time to consider my proposal."

Miles and hours passed by, and all of us were weary when we finally reached the city limits of Muskau. It was past three-thirty. We marched two more miles through this city of thirty thousand to the industrial area before we were stopped. Pop-Eye held an appell, and he was really upset when he discovered that five more of his guards had disappeared. All prisoners were accounted for. He called all the guards of the Center Compound together, and told them that he would shoot the next guard he caught leaving the formation for any reason whatsoever.

The German Civilian Authorities of Muskau had been notified by the commandant earlier that morning that they would be expected to furnish shelter for eight thousand American prisoners of war. We were marched to the edge of an industrial area and assigned to abandoned factories and plant buildings. Many factory workers were being evacuated, and labor shortages made it necessary to close many small plants. Our leaders made vigorous protests against being housed in defense plants and factories, because of Allied air raids, but there were no other quarters available. Many Kriegies were alarmed over our quarters. They knew that if the city were bombed, the factories would be the center of the target. However, most of us were too tired to care, and many men calmed down when they discovered the small city had not been bombed since the beginning of the war.

The Center Compound was quartered in a brick factory which was two hundred yards square and three stories high. The factory had been closed due to material and labor shortages, and most of the machinery had been removed. It was heated by a large furnace in the basement. Colonel Spivey traded the night watchman ten packs of cigarettes for a small wagon-load of coal, and the furnace was fired long enough to heat the building. Six blocks and four combines were assigned to each floor. Our block was assigned to the third floor, and Combine C drew a space along the wall in the center of the building. We had one large window in our area. Each combine leader was given a piece of chalk, and the combine areas were marked off so no one would hog more than his proper share of space. There was actually plenty of room for all of us, but Kriegies either had a system for everything or created one on the spur of the occasion. There were toilets and water fountains on every floor, and we found our quarters to be very comfortable.

The physical condition of the compound had improved a great deal, since the upsurge of sickness that had swept through our ranks the second night of the march in the Catholic church at Wharton. All of us had lost several pounds each since the march had started. At least one hundred fifty men were still suffering to some extent from the effects of dysentery and diarrhea. Many men had colds, and some Kriegies still had symptoms of pneumonia. Twenty-two men were still listed as critically sick; however, the overall physical condition of the compound had greatly improved.

The entire five days of marching had been made in below freezing weather. Three of these days of marching had been under the most adverse weather conditions, with bitter snowstorms and long hours of marching. Some five hundred men had become sick before the end of the second day of the march. Two hundred of these men were considered critically ill. We were without doctors or medical supplies. They were all twenty or more pounds underweight, and their resistance was unusually low. Almost any doctor would have advised that to move these critically sick men in freezing weather, and march them in the snow for long hours at a time, would be sudden and certain death for nearly every one of them.

You didn't have to be a doctor to make this summary. Anyone with common sense was aware that the two hundred sick Kriegies of the Center Compound were moving despite all the laws of Nature. Faith was one thing, but the cold, hard facts of life were something else. We hadn't learned too much about faith, but we knew a great deal about the facts of life. According to medical science, these two hundred men had been doomed to die.

Strange as it may sound, ninety-nine per cent of them were still alive after five bitter days of snow and bad weather. The weather, marching, and snow continued to plague them, yet they steadily improved almost to the last man. Snow, boiling water, and hot stew were the miracle drugs, and men from all walks of life took up the practice of medicine.

I had made two great discoveries—the most important discoveries of my entire life. One was the proof that the human body can take many times the punishment I had previously been taught was possible. If the desire and will to live are strong enough, a man can endure unbelievable hardships and survive. Encouraged and challenged by his comrades, a man can do the impossible.

The second discovery was much harder for me to accept and understand. It is the fact that there is a living God moving with us and watching us every hour and every day. I knew that many people never found Him though He was close at hand. I remembered sincere people in my life who had prayed, and their prayers had not been answered. I remembered others who attended church, sang in the choir, and listened to the preacher's sermon, but they had never been able to reach God. I remembered still others who didn't care one way or the other. My amazing discovery was the fact that God is always there, but the key to reaching Him is faith, a belief in Him without question or doubt. One must have faith great enough to drown all fear. I learned that this kind of faith opened the door between God and man, and that He stepped forward to answer the request. I also learned that He created this world and made the laws of Nature, that He could, and sometimes would, set them aside. I discovered that He could and did perform miracles, and that He could make His presence known to those of real faith, without being seen.

These two discoveries gave me the answer to why two hundred dying men had lived and recovered under impossible conditions.

The men of my combine were busy unrolling packs and preparing beds for the night. Each of us had two blankets, one to lie on and one to cover with. Overcoats were used as pillows. Clark sent Dave Roberts and me to store the sled in the basement. When we returned to our third floor quarters Bill Mott and Bob Rhinehart were preparing supper. The rest of the men had gone to the toilets to wash with soap and water. We joined them and washed for the first time in five days. We shaved in cold water and passed several small mirrors up and down the line. We combed our hair, brushed our teeth, and returned to the combine. All of us got into clean underwear and hung our clothes along the wall to air and dry.

We were all too hungry to talk during the evening meal. While we were eating, Captain McGee came by and told us the lights would be turned out at 10:00 P.M. It was just 8:00 P.M., so we knew we would have plenty of time to finish eating and have a bull session.

Matt Johnson took the floor almost at once.

"If there were a chance of our being liberated, the Germans would march us for a solid week, if they had to. They are holding onto us at all costs, and they must have good reason. It wouldn't surprise me at all if they take us straight to Berlin. Maybe Hitler is going to trade the ten thousand of us for his life. Would the U.S. agree to that, if they knew we would die otherwise?"

"I doubt it," Gene said. "I have thought of another maybe. Maybe they are going to take us to Berlin and scatter us all over the city. If the Germans sent a message to Allied Headquarters saying that ten thousand American and British officers were scattered all over Berlin, and that all of us would be put to death if the city were bombed again, what do you think would happen then?"

"The air force wouldn't think of bombing Berlin, if we were held as hostages," Dave Roberts put in.

"I disagree with that, Dave," Ted Carson interrupted. "Our country is out to win this war at all costs, and nothing is going to stop them from bombing Berlin until the goons surrender. We would simply become part of the price of victory."

"I agree with that," Rhinehart finally said. "I agree *if* we are taken to Berlin, but I think the chances are greater that we will never go to Berlin, regardless of some big shot's orders. The Germans in charge of us are just as eager to live as we are. If we go to Berlin they will have to go with us, and if we die they may die also. All German officers, except the Nazi party fanatics, know that the war is lost. Those in charge of us are smart enough to know that their treatment of us will determine the outcome of their lives. They also know that Berlin is under attack day and night by the air force, and with the German army in full retreat, almost all communications are out of order. Hitler couldn't touch them if they disobeyed orders now."

"That is my theory exactly," Byron Clark said. "Why should they do anything to harm us when we are their best chance of staying alive? The commandant has sense enough to know that American Intelligence is keeping up with us, and that all Allies have been warned of our daily location. He knows that neither Russian nor American forces will bomb or attack us, unless it is an accident. His best chance is to deliver us to a safe place and let us be liberated."

Bill Mott said: "Then why hasn't he let the Russians capture us?"

"Because," Clark answered, "he has had the S.S. on his neck ever since we left Sagan. He has got to get someone higher up to help him, or he will have to get rid of the S.S. If he tried to disobey orders now, they would arrest him and take over themselves."

The lights blinked, and our meeting broke up. Each man found his bed and crawled in.

When I opened my eyes again, it was Thursday morning, February 2, 1945.

"Come to the window," Mott shouted.

Some Kriegies were outside getting water and washing for breakfast. German guards were standing about watching them, but the surprising thing was the large number of German civilians. There must have been at least two hundred of them on our side of the building. Word certainly got around in a hurry. They had potatoes, brown flour, vegetables of all descriptions, several pots of boiling water, and small amounts of meat. The goons were complaining about building fires and heating pots of water next to the building, but the civilians laughed at them. Some politician was serving free coffee to the guards. They all started smiling when they got free coffee, and we figured this guy was the broker. He probably set up the whole trading post and got a commission on the deal.

Mott and Rhinehart got dressed as fast as they could, and Pirtle insisted on going with them. Some German fraulein shouted in English: "Come down and trade." A number of Kriegies misinterpreted her invitation, and the stairs going down were jammed. Captain Randolph would only allow two men per combine to leave the building at a time, so Mott rejoined us at the window.

We lost Pirtle and Rhinehart in the crowd, but when they returned they were loaded with vegetables. Pirtle had appointed Rhinehart as grocery boy, while he personally made every purchase.

"It's all in knowing how," Pirtle explained. "Rhinehart carried the groceries, and I did the trading. Never offer to trade anything for a full pack of cigarettes. I opened one pack at a time, and I never made a trade for over five cigarettes. There are enough vegetables here to make a stew for several meals, and it only cost us forty-eight cigarettes."

Mott and Rhinehart boiled the stew in our dishpan, and we ate it for breakfast, dinner and supper.

Shortly after breakfast, the Germans announced that we would stay at Muskau for the rest of the day and another night.

"That settles it," Tom explained to us. "The Russians have been stopped dead in their tracks. In all probability they are retreating. They never were good fighters anyway, and the commandant wouldn't be issuing such crazy orders unless the Russian lines were collapsing. Swanson, you idiot, where are those Russian tanks you were telling me about last night?"

Later in the morning Colonel Spivey sent a member of his staff to every combine. We were told that General Vandermann was meeting with the commandant and a general in the Luftwaffe. They were discussing terms of the Geneva Convention and our destination. When the meeting was over, the general would issue a statement.

This news created much excitement, and was the main topic of conversation the rest of the day.

In the afternoon, eight of the large horse-drawn wagons were emptied, and we were issued one Red Cross food parcel to every four men. Each combine got three new parcels to replenish its dwindling food supply. Each combine member was issued a package of cigarettes from the parcels, and the rest of them were kept for trading purposes. We also had new supplies of soap and chocolate, from the parcels, for future trading. So far, our cigarettes had held up well. Most of us had a large reserve in our personal packs, from months of saving.

Everybody was waiting for the report that was expected from the general. Pirtle got out the cards and started a poker game. The rest of us slept.

After supper, Carson and Wynn were washing dishes when we finally got the general's report. When Captain Randolph reached us he sat down on the floor and said: "Do not shout or make any comments while I am reading this report. It is from the general, and it reads as follows: 'Gentlemen, we have not had a news report for several days, so I will try to bring you up to date. Sagan fell to the Russians on Sunday, January 29th. There has been heavy fighting ever since the fall of Sagan, and Wharton finally fell into Russian hands today. The Russians are twenty-three miles from us, but they have been stopped by a German counter-offensive. In some places they have lost minor ground. It appears they will mop up the present part of Germany under their control before advancing further. However, you must remember that one major thrust could liberate us at any moment.

"'At dawn this morning seven Allied armies launched an all-out attack on the western front. The entire Allied air force was used to support ground troops. Allied headquarters said it was the final major offensive of the war, and that our troops would advance across Germany until they meet with the Russian army. No advances have been announced.

"'For several hours this morning and this afternoon I met with the commandant and a general on the staff of the Luftwaffe. Under the direct orders of Hitler, the German air force has been ordered to march us to Berlin. We are to be held in the city as hostages to prevent further bombings from the Allied air force.

"'I have explained to the general that this would be considered direct and willful murder of prisoners of war. I have also explained to him that we have no intention of going to Berlin, and that if anything should happen to us, he and every general in the German air force will be held as directly responsible as if they had issued the orders themselves. I have made him understand that the war will be over in a very short time, and that Hitler is doomed. I reminded him that any attempt to march us to Berlin would automatically sign his and many other German generals' death warrant.

"'We will leave here tomorrow or the next day for Spremberg. There will be another meeting there, after the general confers with his superiors. I wouldn't worry too much about this news, because I have made a proposal that will be given serious consideration.

"'At Spremberg, I intend to change their minds. If for some reason I should fail, and if the Russians are within twenty miles of us, I will probably order you to attack the guards and hold out in the city. The guards could easily be overpowered, and we could hold out or hide out for several days. This will be done only as a last resort. Under no circumstances will any man desert or make any foolish attempt on his own, unless I issue the order.'"

Everybody became excited, and theories and predictions were discussed all over the third floor of the brick factory.

The next day we waited for orders to resume the march. Shortly after ten o'clock, we were informed that the commandant was in a high-level conference with officials of the Luftwaffe, and that we would stay in Muskau the rest of the day and night. The commandant had received orders to start marching us to Spremberg the following morning at eight.

The brick factory covered two city blocks, including the outside grounds used for storing and loading bricks. Pop-Eye moved his guards back to the outer edges of the factory area, and allowed us freedom of the grounds. Most of us wanted to walk about and exercise our limbs. Pop-Eye was agreeable to this, for he wanted us to keep in good shape for the march. It was still freezing, and snow was packed heavily over the ground; however, the sky was clear, and the sun beamed down on us.

Several hundred planes flew over during the day. Some were Russian bombers flying toward Berlin. Others were German fighters, flying toward the front lines. We stayed outside as long as possible, on the chance that we might see an air battle between Russian bombers and German fighters. Some battles must have occurred, but Muskau never saw any action while we were there.

Half of the men in my compound were asleep at one time or another. Some Kriegies took mid-morning naps, while others slept in the afternoon. Most of us knew that sleep would give us added strength, and it was the general feeling among us that this march was far from over.

That night the news was read at 9:30. We had a new reader, but we didn't bother wasting time with introductions. He started reading immediately.

"We will depart for Spremberg at 8:00 A.M. in the morning. We should arrive sometime tomorrow evening. Quarters are being prepared for us to stay in Spremberg for at least one night.

"The American 1st, 9th, and 3rd Armies are driving across Germany toward the Rhine River. All ground lost in the Battle of the Bulge has been recovered with additional gains along the entire front. Heavy fighting is in progress. The Siegfried Line and the Rhine River are the last two main barriers between Allied front lines and Berlin.

"The Russians have been stopped on the eastern front, and are still reported mopping up and bringing up reinforcements."

Saturday morning, February 4, 1945, we were up early rolling packs and loading sleds. The two days' rest at Muskau had been of great benefit. Pop-Eye was in a much better humor than he had been since the beginning of the march. Pop-Eye was now holding two appells: one to count us, and the other to count his guards. He was very pleased to find all guards and prisoners accounted for.

When we left the city limits of Muskau, it was nine A.M. The sky was still clear, and the sun was out. All of us were in good spirits, and we were hoping there would be no more snowstorms.

The highway to Spremberg was much wider than the highways we had previously traveled, but the heavy traffic was the same. The greater highway width didn't help because there were many more civilians traveling with us than during the first days. We decided that most of the inhabitants of northern Germany were on the march. It was still cold, but the weather was pleasant compared to the previous snowstorms.

We passed through several small German towns, and made four fifteen-minute stops during the day. It was starting to get dark when we reached Spremberg. There were several dozen large farms at the edge of the city, and we were housed in barns for the night. Our barn quarters were similar to those at the farm village between Wharton and Muskau.

German housewives and frauleins were waiting for us at the barns with hot water and vegetables to trade for cigarettes, chocolate, and soap. It always amazed me how they managed to spread the word so fast. The German guards always cooperated, because as long as we were in good marching condition their jobs were much easier. They had plenty of food, but they were short on coffee and cigarettes. The compound leaders always took care of them to keep them in a cooperative spirit.

After supper, we waited for the news. Everybody was tired from the long march, and most men talked from their beds.

Captain Randolph brought the news report just past nine o'clock. "We were unable to get the war news tonight. Sometime tomorrow I will meet with several generals of the German air force at Division Luftwaffe Headquarters here in the city of Spremberg. I ask that each of you pray for me tonight. I have also instructed the Chaplains to hold prayer services in the morning. Our destiny will be decided in these meetings. Our future is now in the hands of God." Signed General Vandermann.

XXI

Hell in a Cattle Car

ON SUNDAY MORNING, FEBRUARY 5, 1945 AT NINE O'CLOCK, AN impressive service of worship was held just back of the barns in the small farming community on the outskirts of Spremberg, Germany.

Chaplain Daniel spoke to us with confidence and assurance. He reviewed our entire prison life and reminded us that God had never failed to help us in time of need. He expressed his sincere belief that God would take care of us, even in Berlin. He pointed out that our destination was not important so long as we had the proper faith. He explained that if we had complete faith, God would stand between us and our enemies. He requested us to kneel in the snow and closed the service with the following prayer. "Dear Lord, this service has been held to thank Thee for the many favors rendered in the past. You have saved us, and we shall try to prove ourselves worthy of that salvation. We now face another serious crisis, and we ask Thy mercy. We know that you are wise, O Lord, and that You will decide what is best. We accept that decision without question. Bless General Vandermann, and give him the courage to fight for his beliefs. He is outnumbered, so we pray that you will stay with him at the conference table today. We make this request in the name of Jesus Christ, Thy Son. Amen."

We returned to the barns in silence. The service had been truly impressive, and we were greatly encouraged.

"Do you think the general will do any good with the Germans in that meeting today?" I asked Byron Clark.

"They know what he wants, and they are meeting with him. I think they will go along with his proposal, whatever it is, if they have a fair chance of getting by with it," Clark said. "They want to save their own necks, and if they can see a way they will gamble against Hitler."

"What proposal could General Vandermann make that would be accepted?" This from Ed Wunn.

"Several," Rhinehart answered. "He could ask them to hold us in Spremberg until the Russians overtake us. In return he could promise them their lives and certain concessions after we have been liberated."

"How could they satisfy the big shots in Berlin?" Matt Johnson asked.

"By telling them the highways were jammed with troops, and we could not be moved. They might have a hard time stalling, and again the Russians might liberate us before they even heard about it."

"Why didn't they hold us at Wharton or Muskau?" Swanson wanted to know.

"Because," Pirtle answered, "they weren't sure the Russians would keep moving, and they didn't have a proposition."

"I think, if the general makes a deal, we will be sent to the western front," Clark reasoned. "In the first place, that would be the safest place for us. It would also assure them of some degree of safety, if their plan was discovered. We would be liberated by the American Army, and they could hide out, if necessary, until we were liberated. I wonder how far General Vandermann could go in some kind of deal like that?"

Monday afternoon we were ordered to a new location and were marched to the western edge of the city. Here was a large Luftwaffe fighter base, with many administrative buildings. We decided this must also be their Division Headquarters. We were housed in several hangars; the facilities were much better than the barns. Each hangar had toilets, running water, and plenty of space.

Monday night Captain Randolph read a short message from the general. "Gentlemen, I have been negotiating with several German generals regarding your future. The negotiations are still under way, but we should reach an agreement early Tuesday morning. Everything is working to our satisfaction."

Tuesday morning, February 7, most of us stayed outside enjoying the sun. It had warmed up considerably, and we tried to take as much exercise as possible.

A large crowd of Kriegies were outside the main administrative building when General Vandermann, the commandant, Colonel Spivey, and three generals of the Luftwaffe came outside. They were all smiling. Colonel Spivey excused himself to talk with Colonel King and Captain Randolph. While the colonel was talking, two staff cars pulled up in front of the building with Luftwaffe privates as chauffeurs. In a few minutes Colonel Spivey rejoined the group, and departed with General Vandermann, the three German generals, and two aides. As they drove off, Captain Randolph passed the word an announcement would be forthcoming.

During the noon hour Captain McGee read a message from General Vandermann. "Gentlemen: I have made an agreement with several high-ranking German generals whereby you will be moved, starting late this evening. Several freight trains will be made available to the Spremberg marshalling yard, and you will be moved by boxcar from Spremberg to the western front. Your new prison camp will be Mooseberg, located about halfway between Munich and Nuremberg. You will be out of danger here, and the American Allied General Staff will be informed of your new location. I regret that I cannot continue with you. Everything cannot be revealed to you at this time. I must go to Berlin with the German generals. I am taking Colonel Spivey with me, and I may not see you again until the end of the war. I have talked to the commandant regarding your treatment, and I am certain that you will be treated well. Good luck to each of you. Until we meet again, remember to keep your faith and trust in God."

Later a message from Colonel Spivey was read. "Men of the Center Compound. I hope you have a very good trip to Mooseberg. I certainly didn't want to go to Berlin, but I had little choice. It is necessary to secure your safety. Colonel King will take my place as Compound Commander. Follow his orders as you have followed mine, and you will have nothing to fear. God bless you until we meet again."

We went wild with excitement. We were indeed happy about going to Mooseberg, but a long trip in boxcars would probably be plenty rough. And we were alarmed about the loss of General Vandermann and Colonel Spivey. Clark summed it up when he told us during lunch: "I would almost rather lose my right arm than the general. Now that he is out of the way, things could get much worse. I have always thought the commandant was rather stupid, and at times he hasn't known what to do. The general has usually had to tell him."

"You are right," Mott agreed. "The commandant will never see the day that he is as smart as Dumbkoff, and Dumbkoff is almost an idiot. I wish I could take Dumbkoff home with me when this thing is over. My dad has a farm in Connecticut, and we could put Dumbkoff out there gathering and counting eggs. Can't you just hear that stupid jackass counting those eggs!" Mott began to laugh, and all of us laughed with him.

That afternoon we were notified to be ready to march at three o'clock. When the appell had been taken, Pop-Eye made a speech.

"We are leaving here to board several hundred boxcars at the Spremberg marshalling yard. From there we will go to Mooseberg, to Stalag Seven-A. The trip will take three days and nights, and we should arrive there Friday at noon, if we are not bombed by your air force. This camp is run by Storm Troopers, and you will learn the meaning of discipline."

When we got to the marshalling yard, we learned that the commandant had received a change in orders whereby four thousand of us were to go to Mooseberg, and four thousand to Nuremberg. The first eighty cars were to stop at Mooseberg, and the remaining eighty cars were to go on to Nuremberg. Every compound and every block was split up, and men were marched to the separate trains. Nine members of my combine were selected to go on the first train to Mooseberg. Peter Glass, Dave Roberts, and Ted Carson were assigned to the Nuremberg train, and we were never to see these men again. We felt as if we were saying goodbye to our own brothers.

Our compound was formed into groups of fifty men. Each group was now marched to a boxcar for loading. But as cars were loaded, open revolt broke out. Most of the cars had been used for hauling cattle. The inside of a cattle car was filthy, and the smell was almost unbearable. Moreover, these cars were about half the size of an American boxcar, yet fifty Kriegies were crowded into each car. There wasn't room for us to lie down, or even sit down. It was very difficult just to try to move about. It was apparent that many men would have to stand for the seventy-two-hour trip, and in our weakened condition it was impossible for most of us to do this.

Several hundred men jumped from the cars refusing to stay inside or move off under such conditions. Then several thousand of us started shouting and sitting on the ground, refusing to board the cars. When news of this reached the commandant he went into a rage.

He appeared with Pop-Eye and six German sergeants armed with submachine guns. He fired his pistol in the air, and demanded we return to the cars. We refused to move. He thereupon proceeded to the front car behind the engine. This car was loaded—after guards took up firing position and fixed bayonets. The commandant was shocked by our open defiance, and he was outraged because his orders were disobeyed. We knew he was crazy enough to order the guards to shoot, if we again refused to carry out his order. After the first car was loaded, he had the doors closed and locked. He proceeded to load each car and to lock the doors. Our departure was delayed for over two hours because he made a personal speech in each car. After our car was loaded, he stood on the platform at the door and told us:

"You have been foolish enough to disobey my orders and threaten me with open revolt. You are now going to be taught a lesson. You will be locked in this car for twenty-four hours. If at the end of that time your confinement without light, air, or water has taught you to obey, then I will consider giving you better treatment. If you have not learned your lesson, then you can stay in these cars the rest of the trip."

As he departed down the gangplank, the doors were closed and locked.

We stood there in the dark shocked and bewildered. For the first time, as prisoners of war we had suffered a major defeat. We were now going to be treated as slaves or criminals.

Except for a few cracks of light, the car was dark. We did our best to accept our surroundings and prepare for the trip. We divided the car in halves, and twenty-five men were assigned to each half. Space was then so allocated that half of us could lie down, and the other half could sit in cramped quarters or stand if they desired. We scheduled a rotation every four hours until the trip was over.

Food packs were stacked against the walls, and individual packs were spread out in the crowded space. Four pasteboard boxes were placed in each of the four corners of the car—to be used for toilets or sickness. Our blankets were used to cover the hay and cow droppings. We didn't have to lie in the offal, but the smell was forever present.

At 8:00 P.M. the Cattle Car Evacuation began. Two trains, comprised of approximately eighty-five cars each, moved slowly out of the Spremberg marshalling yard. The commandant had the civilian railroad authorities in Spremberg attach three chair cars to the rear of each train. The commandant, Pop-Eye, and other German officers, sergeants, and guards prepared themselves for a most pleasant journey in the chair cars.

Inside the cattle cars, we had started a trip that would turn men into swine. The commandant was wrong about the stale air. Cold air gushed about us through holes and cracks in the sides. We kept our overcoats on, pulled our caps down over our ears, and put our hands in overcoat pockets. The drafts of freezing air swept through the car, and most of us were coughing before the night was over.

However, the cold, fresh air did serve one purpose. It reduced the odor from cattle droppings. Men had to move to the pasteboard toilets all during the night, and new, unpleasant odors were created. It was very difficult to use a small pasteboard box amid crowded conditions in a small boxcar that was jerking and swinging constantly. Moving about the car proved impossible without stepping, hitting, or falling on a fellow Kriegie. But in many months we had learned the lesson of tolerance and patience. Men restrained themselves as they were shoved, jerked, and stepped on. We changed the shift every four hours, but it was impossible to sleep. The hours moved slowly by, and the train rocked back and forth. Men became exhausted from the lack of sleep and long hours of standing. By dawn we were worn out, hungry, and thirsty. The smell grew steadily worse, and nerves reached the breaking point.

Shortly after daylight the train stopped, and we could hear people blabbering in German. The German guards were checking our cars to make sure they were still locked. German civilians were talking to them, and we recognized the world "Leipzig" several times.

Food packs were opened, and we ate bread, sandwich spread and cheese. But our mouths and throats were dry, and the food was very difficult to chew or swallow. We gagged and choked, and many men could not finish eating.

"I feel like I am choking," Matt Johnson said. "My lungs hurt, and I am getting short of breath. Do you suppose they ever intend to open these doors and give us any water? We have got to have water, or we will die." This was to be a prophetic statement for poor Matt.

Gene became sick and couldn't even talk. I knew when the puking started, it was really going to get rough. Ed Wunn and Gene were the first two to start, but many followed. This first night in the cattle cars was the beginning of the worst ordeal of our lives. By noon fifty per cent of the men in my car were sick. Dysentery reappeared, and cramping stomachs followed. The odors were enough to make a strong man sick.

The greatest suffering was from thirst. We thought we had experienced every kind of pain, but lack of water was something new to us. Of all the suffering any of us had ever known, being without water was by far the worst. Our tongues began to swell, and our throats felt as if they were closing. We finally got too dry to talk or swallow. Lips were parched and eyes burned. No one could think of smoking a cigarette. We knew it was impossible to eat again without water.

Finally the toilet boxes started overflowing. As men moved away from them we were packed like sardines, jammed one against the other. Every man was able to control his temper with his neighbors, but hate of the Germans mounted as we listened to the car wheels clicking over the endless railroad track.

The rays of light from the outside disappeared. It was getting dark again, and we waited for the train to stop and the car doors to open.

When we finally stopped, we heard the word *Plauen* spoken by several guards. Two guards took up positions at each car, and the doors were opened. Toilet boxes were dumped and thrown from the cars, splattering about the railroad station, as we climbed out of the cars and fell in line for an appell. The guards were shouting and cursing about the boxes being dumped all over the station, but we ignored them. Our hatred for the guards was at an all-time high, and we gave them absolutely no cooperation.

We started shouting "Wasser, Wasser," and when they started taking the count men broke ranks everywhere, walking up and down and defying the guards. Rifles were fired in the air several times, but the Kriegies disregarded the warnings. We were past the stage of caring about German reprisals. The commandant was shocked beyond words by the display of defiance. Three times the appell was taken, and many Kriegies were slapped and struck for falling out of ranks, but the confusion and disorder continued. Pop-Eye never got a total to report.

The commandant sent for Colonel King, and ordered him to make us fall in formation.

Colonel King spoke defiantly: "You will not get an appell until you give us water and clean these filthy cars."

"Unless you order your men to obey me, I will consider that a mutiny has taken place; and I am going to open fire on them," the commandant shouted.

"You open fire, Major," the colonel shot back, "and you won't even get a trial when the war is over in a few days. You and every guard here will die a death of torture that will be remembered forever in German history."

The commandant's face got dark red as he ordered the cars reloaded and locked. The train was delayed for over an hour before we departed. While they were reloading us several civilian officials were shouting and complaining about the appearance and smell of their station. The commandant hurried the guards, and we knew he was relieved to get moving again.

During the night many men grew critically sick. Again it was impossible to sleep. Sickness smells and other car odors were horrible. Many men had bad colds, and some were probably developing pneumonia. Several men in our car became hysterical and started screaming and crying. Calming them down was difficult, but all of us tried to console our neighbors. Nerves were slowly stretched to the breaking point, and men began to revert to the animal stage. We were being treated like animals, and it was easy to see that with much more of this, we would start acting like animals. Thursday morning the train pulled into Regensburg. The doors were opened, and we were unloaded for another appell.

All of us started hollering "Wasser, Wasser." We had made up our minds that we were either going to get some water or die. The station was crowded with civilians trying to board trains. There was a pond just ahead of the engine, and there were water fountains in the station. We broke ranks en masse, and started for the pond and water fountains. Guards fired in the air, but all of us moved to the water. Men drank and filled their cans and jars with water.

Until that moment I had never realized the value of plain water. It was cool, refreshing and wet. It was the most valuable necessity of life. Without water men became animals.

As Kriegies finished inhaling large amounts of water, they moved back across the tracks toward the cars. Many of us stopped along the tracks and used the area for a toilet. Several hundred civilians watched. Women were shocked, but we had reached the point of indifference. The toilet boxes were again dumped from the cars.

Colonel King ordered us to fall in and we complied. The commandant was utterly stunned by the sight before him. Pop-Eye had talked the commandant out of firing on us. While Pop-Eye was taking the appell, three Gestapo officers joined the commandant and severely criticized his lack of control of the prisoners. The station superintendent joined them, and started shouting about our trains causing a traffic jam. Pop-Eye ordered all men to board their cars, and reported that all men were accounted for. We didn't waste a minute reloading. The Gestapo officers were confused by so much disorder, and they were trying to decide who should be punished for this unheard-of conduct. The station superintendent ordered the trains to move, and the commandant gave the signal to depart. The commandant, Pop-Eye and the guards boarded their cars when the trains started moving. The Gestapo officials and the superintendent were still arguing among themselves as we moved out of Regensburg.

It was a great victory for us, and a strong morale-builder. We had obtained water, and this brought life back to most of us.

We decided that the commandant was so shocked he didn't know what to do. We knew that Pop-Eye had talked him out of shooting any of us, and that he probably had reminded the commandant that the war would be over in a few days. Pop-Eye knew the war was lost, and he didn't want an American military court trying him for murder when the war was over.

Soon after the train started, Mott had a brainstorm. He tied two blankets together, and swung a hammock in the far end of the car. Men started looking for places to tie their blankets, and when we rearranged ourselves, we found that eight hammocks were swinging in the car—a great relief to the crowded floor space, for men could sleep in the hammocks. Our combine prepared breakfast, and the water was rationed to each man.

Matt Johnson looked sick, and I asked him: "Matt, how do you feel?"

"I think I have fever. My lungs and throat are so sore I can hardly breathe or swallow."

I felt Matt's head, and indeed he had high fever. Clark helped me fix his pack, and we supported him to Matt's hammock. He went to sleep almost immediately. Most of the men in our car were sick, and I could see no hope for improvement until we got out of the cars at Mooseberg. I helped Clark, Rhinehart, and Matt clean up the dishes and food.

"This is the way I've got it figured," Mott said. "Did you notice that the snow is beginning to melt? It is getting much warmer. When we get out of these damn cars, life should be much better. One of these days the gates of Mooseberg will crumble before American tanks. Pop-Eye's glass eye will probably pop out, and Dumbkoff will be hiding under my bed. I wonder if Dumbkoff could ever learn to speak good English?"

We talked for some time, and every now and then Mott would give us a good laugh. Later in the morning all of us joined the sleepers for a nap. I was so exhausted from lack of sleep that the hard boxcar floor didn't bother me a bit. I went to sleep along with the others.

Late Thursday afternoon we arrived in Munich. The snow had melted, and it was a great deal warmer. We fell out for an appell, and all of us cooperated with Pop-Eye. The commandant was delighted to find everyone carrying out his orders. The sickest men were allowed to remain in the cars. Pop-Eye told us the doors would not be locked the rest of the way, and that we could open them if we desired fresh air. We were allowed additional water, and we walked up and down the line of cars taking exercise. It was nearly dark when we were reloaded for departure.

Gene looked worse than I had ever seen him. None of us had shaved or bathed for a week, but Gene's face was sunken, and he had dark circles under his eyes. He had become extremely bitter.

Pirtle was feeling bitter also. "Hell, I am so sick, I would have to get better to die. When you think about it, dying isn't so bad. At least it would be an end to this filth. What if we struggle on and on, and then in the final hours before we are liberated, the Germans shoot all of us."

"The general got those German big shots to move us to this area," Byron Clark interrupted. "He wanted to get us as far away from Hitler as he could. No German officer is going to carry out any order to shoot us unless he wants to die himself. Most of these German officials have accepted defeat, and they want to end this as bad as we want to. They know that our survival is the only way for them to stay alive."

The train finally pulled out of Munich. Matt Johnson was still coughing and turning in his hammock. We had fed him water several times. He drank the water, but he was delirious. He mumbled to himself whenever he was awake, and didn't recognize any of us.

It was sometime after midnight when I went to sleep, and when I woke up the next morning the train had stopped.

We were unloaded on the north side of the city of Mooseberg. About half a mile beyond us we saw another large prison spread over the ground.

It was eleven A.M. on Friday, February 10, 1945. The march of death had come to an end. We had traveled across a large part of Germany, a distance of 480 miles. It had taken us thirteen days to make the trip, and we had suffered many casualties. This march would be recorded as part of the history of World War II. Many people would soon forget, but for us it would live forever. We had survived the most horrible experience of our lives, and we had gained knowledge that was priceless.

XXII

The Latrine Revolt

LEAVING THE CATTLE CARS, FOUR THOUSAND SICK KRIEGIES FELL IN block formation for an official appell. Some men could not stand, and many were supporting the men next to them. Over three thousand men were sick with infected stomachs, dysentery, colds, and pneumonia. All of us were weakened from the effects of malnutrition and mental and physical exhaustion.

When the appell was completed we started marching the last few steps to the gates of Stalag Seven-A. When the front columns reached the gate, the formation was halted while our commandant talked to the new commandant of Mooseberg.

I turned just in time to see Matt Johnson collapse in a heap. Several of us rushed to him, and Clark lifted his head and shoulders from the ground. Pirtle wrapped a blanket around him and loosened his shirt.

Matt was breathing deeply. His pulse was very weak, and his heart was skipping beats. He opened his eyes and looked at us as we knelt about him. He looked up into the clear blue sky and spoke slowly: "Isn't it beautiful," he muttered.

We looked into the sky and then down at Matt. His head dropped in Clark's arms. He was dead.

We covered Matt with the blanket. Pop-Eye sent two guards and a wagon to us. Matt was loaded on the wagon, and Pop-Eye said he would be buried. In a few minutes the formation started marching through the gates of Mooseberg.

Stalag Seven-A extended across acres of ground. It was built in compounds like Sagan, each enclosed in a separate barricade. The barbed-wire fences encircled the camp, and the guards and sentry towers were everywhere. The camp wasn't much larger than Sagan, but there were many thousand more prisoners. There were several American Compounds, one British Compound, one Russian, one French, and one with soldiers of all Allied nations.

As we moved along the streets, prisoners stood near the fences watching us. Some of them hadn't had a bath or shave in days. They weren't being fed well, and most of them looked weak and sick.

We stopped again in front of the Center American Compound. Our prison identification cards were moved in a wagon to the front gate. Each man's prison tags were checked. A German guard called each Kriegie's prison number from the tags, and his identification card was pulled and placed in a new box. When each man was identified, he was moved inside the gate.

The Center Compound of Stalag Seven-A was surrounded by three adjoining compounds. An American compound joined us on the northeast, the British compound joined us on our west, and the French compound started at our southern boundary.

Each compound had ten barracks, two latrines, and one small kitchen for issuing rations and water. There were more compounds than at Sagan, but each compound was much smaller.

The barracks were built for one hundred men; however, three hundred or more men were housed in each. There were no partitions between the combines, beds were closer together, and they were built three tiers deep. After space was set aside for kitchens, storing food boxes, and personal belongings, there wasn't room to turn around. Three combines had to share each table, and the eating schedule was arranged in shifts. Games were also arranged by shifts, as there were three times as many men at each table as we had previously had at Sagan, and the tables were occupied three times as long during the meals.

The commandant and commanding German officer at Mooseberg was a colonel-general. Originally he and his staff of German army officers and guards made up the personnel of the camp. As more parts of Germany were captured, more and more prisoners were evacuated to his camp. When we arrived on February 10, 1945, there were 35,000 prisoners at Mooseberg, and the camp was being operated in sections with assistant commandants from all branches of the German armed forces. Our commandant was an assistant commandant in charge of the Center American Compound. The guards were now German army and Luftwaffe. The German army general was commandant of the entire camp, but he had supervisors from the S.S. and Gestapo visiting constantly.

The S.S. and Gestapo did not believe in the Geneva Convention, and they hated all prisoners of war. Especially, they hated American fliers who had bombed and destroyed cities of Nazi Germany. As American, British, French, and Russian armies advanced daily across Germany, their hatred increased.

German army and air force generals battled constantly to keep them from taking control of their camps. The commandant was rather fanatical himself, but he would not allow them to go too far.

The first week after we arrived at Stalag 7-A we were busy cleaning and arranging our quarters. The first several days things were badly disorganized, for most of us were too sick to think of doing anything. We drank large amounts of hot water, ate mild rations, and slept most of the time.

After several days' rest most of us were able to start to work. Barracks were cleaned with cold water, soap, and rags. Combines were reorganized, and cooking, cleaning, and eating schedules were set up. Sick men were cared for by those of us who were able to work.

Water was rationed, so it was difficult for us to take any kind of bath. When we could get several tubs of spare water, twenty men would take washrag baths from the same tub of water. Some times you couldn't get very clean, but at least the water was refreshing. We were able to shave about once a week.

Hot showers were not given at Mooseberg, and coal was rationed to almost nothing. It became more difficult for the Red Cross to get food from Switzerland to Mooseberg, and parcel rations were cut in half. The Germans were losing territory every day, and their transportation system was under constant bombing and strafing attacks. Our German food rations were cut to one weak bowl of soup and two slices of bread per day.

At Mooseberg our daily activities were changed a great deal. There was no library, auditorium, or equipment for outdoor games. We spent hours of our spare time walking, sitting in groups, exercising and talking.

On Saturday, February 18, we had our first real warm day since we left Sagan. Immediately following the inspection, we got our first news of the war and of General Vandermann since we left Spremberg.

Colonel King read the report.

"You had a good inspection this morning and I have a few important items to read you.

"First of all, we all realize that we have many sick men among us. Our food rations have been cut in half, and we are steadily losing weight. I urge you once again to get as much rest as possible, and keep your spirits high. Each of you should do your best to cheer up and help your neighbor. The war can't last much longer, and we will be out of this hole.

"We will start giving you the news again. I think the goons know we are getting it, but they don't seem to care. It is all bad to them. Rhoder arrived here this morning and reported to the commandant. He went to Berlin with General Vandermann, Colonel Spivey, and the three German Luftwaffe generals. They gathered certain papers, money, and belongings and returned to Spremberg. Rhoder said the city of Berlin was in flames and under tremendous aerial bombardment. From Spremberg, General Vandermann, Colonel Spivey, three Luftwaffe generals, and Rhoder drove to Munich. At Munich, Rhoder left the two staff cars, and came here to report to the commandant.

"The rest of the party drove to the German-Swiss border and crossed into Switzerland. Naturally the border guards never questioned three German generals. Rhoder reported that General Vandermann had promised freedom from war crimes trials and return to satisfactory civilian life of the generals and those guards who did not harm American prisoners of war. Rhoder stated that the general was taking the German generals to Supreme Allied Headquarters to report on the war and our location and condition. General Vandermann will also report on our needs, so plans can be made to evacuate us when we are liberated. Rhoder was glad to give me the information, since he desires to be on our list of good Germans when the war is over."

"The biggest barrier against our liberation is getting our armies across the Rhine," Ed Wunn commented. "When our armies can get across the Rhine, we will see some real progress."

"I think you are right," Clark agreed. "The Germans will make it their last stand. If and when our armies cross the Rhine, the Germans might as well quit."

Mott was field-stripping a cigarette, placing the tobacco in his small tobacco can. Since we had arrived at Mooseberg, all of us had made small tobacco cans from our food cans. Our cigarettes had been cut in half along with the food parcel cut. Thirty cigarettes had to last a week, and every ounce of tobacco was saved. With four cigarettes a day to smoke, we would only take two or three drags during one smoke. The cigarettes were carefully extinguished and replaced in the tobacco can. A cigarette was smoked several times, and when it got too short to smoke, the unburned tobacco was removed and placed in the can. When we had a can full of loose tobacco several cigarettes were rolled from toilet paper.

Mott put his tobacco can in his pocket and said: "These light rations are sure making me dizzy. Every time I bend over, I start blacking out."

A group of B-26 medium bombers approached the camp from the west. It was against German regulations for us to cheer. But as they got nearly overhead we stood up, cheered, and hollered. Pop-Eye rarely came to our compound except for the morning and evening appell. In a few minutes several bombers dipped their wings and loud cheers roared over the Center Compound. Now Pop-Eye came through the main gate with Dumbkoff, Shisenkoff, and eight other guards. He made us go inside the barracks, and he closed the doors and blackout shutters outside the windows. We had to stay inside for one hour as punishment for disobeying orders.

When Pop-Eye was outside and saw American planes approaching he usually started for the Center Compound. We always cheered, and he always got mad and locked us inside one hour for our punishment.

"Very funny, lieutenants," he told us. "If the Gestapo comes down here, you won't think it is so funny."

The planes were always wonderful to see, and they helped our morale.

Sunday morning, we held our first church service at Mooseberg. The weather was pleasantly warm and the service was held outside in the center of the parade ground.

Chaplain Daniel stood on two wooden boxes near the center of the parade ground. He was smiling, as always, with his Bible under his arm. When all of us had gathered before him he said:

"Let us pray. Dear Heavenly Father, we give thanks to Thee this Sabbath day. We thank Thee for the warm, beautiful weather. We thank Thee for this food and shelter at Mooseberg, Germany. We thank Thee for our health and for our lives. We thank Thee for caring for us on the long march of death and guiding us safely to the western front. We ask Thee to continue to watch over us, and save us from any dangers that may come our way. We pray that this war may be brought to a hasty conclusion and that we may return to our homes and loved ones. Amen.

"Most of our troubles are near an end. We have survived the many months of prison life from the time of capture through our stay at Sagan. We have just recently completed and survived a death march across Germany under bitter weather conditions and at times without food or water. Prison life here at Mooseberg is certainly not as enjoyable or pleasant as that at Sagan, but you should always remember that we are being cared for, and we have the bare necessities of life. Considering the present stage of the war being waged all over Germany, we are indeed fortunate to have anything to eat and any place to sleep.

"Now we must continue the struggle for survival. We must keep our faith and confidence in God. We must believe, if we expect His help. Many of you are wondering if we will run out of food, or if we will be tortured, or killed in a battle over the capture of this camp. I think the record of our past experiences proves that we have been protected and provided for in every major crisis. If we continue that faith, we will be provided for; and if we should ever lose that faith, we may lose our lives as well."

We closed by singing *America*, and spent the rest of the day reading our Bibles and talking about the war.

During the next week, Ed Wunn developed an infected strep throat, and ran a high fever. He gargled with hot water several times a day, but it did not kill the infection. Captain McGee finally requested the commandant to move him to the hospital for several days. The commandant, however, claimed the hospital was overcrowded with strep throats, and that Wunn could not be moved.

The Germans were running out of medical supplies, and it was impossible to obtain more. The Red Cross was sending what supplies it could get through to Mooseberg, and the Germans were using them.

All of us started gargling with hot water to prevent catching the throat infection. I am sure this helped to some degree, because it did not spread to the epidemic stage.

If a sick Kriegie went into a coma, he was removed to the hospital. On Sunday afternoon Pop-Eye at last moved Wunn to the hospital. We heard that night that Captain Randolph had also been taken to the hospital.

Tuesday afternoon Pop-Eye brought us the word: Captain Randolph and Lieutenant Ed Wunn had both died Monday night. It was shocking and saddening news to the entire compound.

I read from my Bible for several hours trying to find words to understand these deaths. I was beginning to think that maybe each of us had a purpose in life. That we had a certain thing to do and the opportunity of so much time to do it. I also realized that some men saw and lived much more than others. I finally put my Bible up, and went to sleep dreaming about our liberation.

The first day of March came, and we were still prisoners of war. When we started the evacuation of Sagan nearly all Kriegies, including the pessimistic ones, had believed that we would be liberated within several days, at the very most. From week to week after that, we had been convinced that the war would end any day.

On this date the closest American army to Mooseberg was 250 miles away, and we still hadn't made a crossing over the Rhine. I swore I would not get excited again until the Rhine was crossed by American forces. After that, I would measure my excitement by the distance of American forces from Mooseberg.

After appell, Colonel King filed a list of complaints with Pop-Eye, and among them he listed insufficient water, cramped living quarters, no provision for hot baths or showers, improper facilities for sick or dying men, and overflowing latrines that were not emptied regularly. Colonel King insisted that he be given an answer from the commandant. He got his answer one hour later. All complaints were rejected flatly, with no reason given.

The west latrine of the Center Compound started overflowing that morning and spread ten feet over the ground on the outside. The smell was terrible, and it increased steadily as the latrine continued overflowing. All members of the compound started using the east latrine until Colonel King stopped us.

He instructed Captain McGee and other block commanders that he wanted every member of the compound to continue using the west latrine until it had overflowed into a circle forty feet wide around the outside. He offered no explanation for this order.

Everybody complained about having to walk through the vile stuff in order to get to the inside urinals. We started holding our noses ten feet from the door, and never took another breath until we were safely outside.

The east latrine was not closed for fear of drawing attention from the guards; however, two Kriegies were stationed inside to make sure everyone complied with Colonel King's order to use the west one.

We were all anxious to know what the Colonel had up his sleeve, and we found out just before the appell. We were told that blocks five, six, seven, and eight were to form at the north, east, south, and west sides of the west latrine for the appell. We were to fall in eight paces from the edge of the building. No matter what Pop-Eye tried to do, we were to stay in that position during the appell.

Pop-Eye came through the gate promptly at five o'clock with his entire staff of guards. Rhoder was out in front of the guard formation talking to Pop-Eye. When the bugle sounded, the block commanders of blocks five, six, seven, and eight waded to their proper positions and ordered their blocks to fall in. All other blocks fell in formation in front of the barracks.

The hundred members of block seven waded to the south side of the west latrine and took up their proper positions. All of us had our pants legs rolled up.

Pop-Eye cocked his eye, trying to figure what we were up to this time. He and Rhoder watched us and then it dawned on them. Pop-Eye started shouting at Colonel King.

"Move those four blocks out of that mess so we can take the appell. What are they doing over there anyway?"

"That is the new position I have assigned them, and they are not going to move. You can count them just fine where they are located."

Pop-Eye was furious. He rushed to the outer edges of the pool and shouted at Captain McGee and the other block commanders to move their blocks forward fifteen paces. They refused to obey his orders. He looked at his watch and realized the assistant commandant would be there in a few minutes to receive the official count.

He ordered Dumbkoff, Shisenkoff, Bear-Face, and five other guards to move through the filth and take the count. When Shisenkoff protested, Pop-Eye screamed at him and he ran through the pool at double time.

As Dumbkoff and Shisenkoff started toward us McGee said: "All right, men, foul up the count until Pop-Eye has to come over here."

After two counts, block seven was five men out of balance. Dumbkoff was scratching his head and Pop-Eye was screaming at him. Blocks five, six, and eight were also out of balance.

The commandant had arrived and Pop-Eye was cursing a blue streak. When the commandant grew impatient, Pop-Eye pulled his billy-club and waded through. As he reached each block, the block was called to attention and the proper count was taken. He was so mad at block seven that he continued to shout as Dumbkoff counted: "Swine, luft gangsters, idiots, swine, idiots, luft gangsters."

The commandant nearly choked from the smell of Pop-Eye's boots when he made the official report. Pop-Eye explained what had happened and the commandant started pounding his fists as he shouted orders. We were immediately ordered to the barracks and locked in for the night. Every man removed his shoes at the door, and our coffee water was used to clean our shoes.

We laughed all evening long for this was the greatest morale booster we had had in weeks.

The next morning we stood the appell in the same position. Pop-Eye brought German police dogs and extra guards with him. He made all kinds of threats to Colonel King, but the Colonel would not budge an inch. During the night the ground had gotten sticky, and it was harder to walk through.

Pop-Eye finally told Colonel King that he would not take the appell until we moved out of the stink. The smell was getting stronger, but we could stand it as long as Pop-Eye could. The commandant arrived, and had a chair brought for him to sit in.

We watched Pop-Eye and the commandant talk and laugh as we stood in the stink. Every time the commandant laughed, we all laughed. He laughed twenty times, and all of us laughed with him. After an hour and a half he started getting nervous. He began walking around and looking at his watch. After another twenty minutes he got red in the face and ordered Pop-Eye to take the count. All the guards waded out to us and took the count. We fouled it up every time until Pop-Eye came storming out with his billy-club.

The commandant left in a rage, and Pop-Eye went to the kitchen to wash off his boots.

That afternoon at two o'clock two sanitary wagons pulled through the front gate and stopped at the west latrine. They had to make two loads each before the latrine was empty. When they left, Colonel King sent twenty Kriegies with buckets and mops to wash the latrine floors. I was one of those twenty men, and each of us was given two gallons of water to bathe with. That evening the appell was over in ten minutes, and Colonel King personally thanked the commandant.

Our status as prisoners of war was somewhat unusual. The large majority of all German military personnel were not fanatical Nazis. They knew that Germany was hopelessly defeated, and they wanted the war to end. We were their prisoners today, but tomorrow they would become our prisoners. They wanted to show their authority, but most of them were actually afraid to harm any of us, because of reprisals. I feel certain that if they had been winning the war, many of us would have been shot long before the "Latrine Revolt."

After evening appell on March 7, Captain McGee called me to his room, "Come in, Ken, and have a seat," he said. "Colonel King has instructed us to reassemble the Camp Record. He has established a safe hiding place, and he wants the project completed tomorrow. Have you got your part available?"

"Yes, sir. It's in my shoe. It is worn and has a slight odor, but I can interpret any part you desire. It is funny in a way, Captain. On the march when I thought I was going to die, I never took off my socks. Something kept telling me that no one else was supposed to know. When we departed for Muskau, I got over the worst part of my sickness, and I recalled the camp record and your lectures. When we stopped for the first break, I took my shoe off and felt inside my sock to be sure it was still there. For the next four hours, I repeated the report several times, and to my utter surprise, I knew it almost to the letter . . .

"Captain, I was really sorry to hear about the death of Captain Randolph," I told him, as I removed my shoe. I handed him the three sheets of crumpled paper. He unfolded them and studied them carefully. He walked to a shelf against the wall, and placed them inside his Bible.

"Randolph," he said, "was one of the outstanding men in our compound. He was sick when he got here, and that strep throat was too much for him. I talked to him before Pop-Eye removed him to the hospital. I was there when they came after him. He knew he was dying.

"Do you know what he said to me? He was propped up in bed, his eyes were glassy, and he acted like he was fighting to keep his memory. He looked at me for several minutes and finally whispered: 'Sam, look after these boys. They are going to need you before this thing is over. I am dying. The good Lord agreed to give me an hour or two more, and my time is just about up. My final prayer is for our men. They will be saved very shortly, but I pray they will use their lives well. I pray that this life as prisoners of war will give them the courage, faith, and determination to do what God expects of them. He is expecting great things of our men, and I want you to tell them this.'"

Shortly before nine o'clock Captain Ferrell appeared with the news. "Men, we have great news tonight, so please listen carefully. First, a message from Colonel King.

"Gentlemen, at last we have entered the final stages of the war in Europe. It has been long and difficult, but now we can plainly see the end is near. I want you to have the highest possible morale, but we must maintain good discipline. In the final liberation of this camp we may be called upon to use our knowledge and skill. You have done a magnificent job. Keep your faith, and remember God in your prayers.

"On March 1, the First American Army commanded by General Hodges, and the Ninth Army under the command of General Simpson launched an all-out offensive in their section of the western front. Treves, Germany, and Muenchen-Gladbach surrendered before the day's fighting was over. Accompanied by heavy aerial bombardment of German front lines, both armies raced forward and encircled the city of Cologne. Several American armored divisions by-passed the university city of Bonn and launched a furious attack upon German forces retreating across the Rhine. On March 6 American armored divisions swept down the west banks of the Rhine and captured the Ludendorff railway bridge over the Rhine River. German troops were utterly demoralized and taken by surprise. Thousands of American troops, tanks and heavy equipment were rushed across the bridge and Remagen, Germany, was captured on the east bank of the Rhine. Several hundred American tanks made their way up the east bank of the Rhine, and engineers launched prefabricated bridges to the west banks of the Rhine in several places. Several additional bridgeheads have been established. German troops were unable to regroup, and were forced to retreat several miles. The Allied lines are now 214 miles from Mooseberg."

Our armies were still over 200 miles away, but the last major obstacle had been surmounted. Now it would be a day-by-day process of smashing German resistance between the east banks of the Rhine and Mooseberg.

We were at the point of hysteria when we went to sleep, dreaming of American troops marching toward Mooseberg.

XXIII

The Sweat-Out

ON SUNDAY, MARCH 12, EVERY ABLE-BODIED MEMBER OF THE Center American Compound attended church. The American armies had been able to advance only an additional five miles since the crossing of the Rhine.

Chaplain Daniel preached a wonderful sermon about our forthcoming liberation. All through the sermon he gave examples of faith. He pictured the end of the war and our return to the United States. He left little doubt in anyone's mind that we were going home in a few days. He always saw the bright side of everything, and it was inspiring to hear him.

That Sunday afternoon was almost like a summer day. After lunch Combine C gathered outside by the east side of the barracks. Nearly every Kriegie in the Center Compound took advantage of the sun, and there was hardly room to sit down.

In mid-afternoon we heard bomber engines in the distance. Then we could see the bombers flying steadily toward us. When they got closer, we identified them as B-26 medium bombers returning from a mission somewhere in Germany. Smoke was pouring from the front of one of the lead planes. As they passed overhead, everybody around us stood up and started cheering and waving. Several bombers dipped their wings, and we cheered again. Pop-Eye was already in the compound with his guards and moving Kriegies to their barracks. Before we went inside, we took another look at the wounded bomber, and observed that she was still holding altitude with her one good engine. Pop-Eye locked us in, and had the blackout shutters closed.

He was determined to break us of cheering, but he never succeeded.

"Did you guys know that I almost had Dumbkoff cheering?" Mott laughed. "I showed him the planes and cheered. He looked at them, smiled, and started to cheer; then he saw Pop-Eye. How can you ever teach Dumbkoff anything as long as Pop-Eye is around?"

That night after a poker game, Captain Ferrell came in to read the news. He had his shirt off, and he had lost so much weight you could count his ribs. He smiled as he took his seat, and we hoped the news would be good.

"Fellows, the news is somewhat better, so this ought to cheer you up. The Russians have captured several more important industrial cities near Dresden, and are now within 48 miles of Berlin in two sectors.

"General George S. Patton, Jr., and his Third army smashed through strong German defenses in the Saarland and encircled several German divisions. The rest of his army is advancing toward Worms on the Rhine. Several armored spearheads have cut off additional German forces between the Saarland and Worms.

"Allied Headquarters reported the First and Ninth armies have established dozens of additional bridgeheads across the Rhine, and that a large number of reinforcements are being rushed to the front lines. That's it."

During the next week, the Red Cross food parcels did not come through as we had hoped. A large part of our parcels were shipped by train, and nearly all German trains were being strafed by American fighter planes.

The International Red Cross started sending truck convoys through. Each truck was white with green crosses painted on both sides and on the top. Because of transportation difficulties, the Germans limited the number of trucks. The small truck convoys worked around the clock, but it was impossible to supply sufficient food to one hundred thousand American prisoners in our area of Germany.

On March 20, our food parcels were cut once again. It was now three parcels to the combine of twelve men. That meant about thirty pounds of food for twelve men to live on for a week. Two and one-half pounds of food per man each week wasn't much to eat.

When the parcels were issued, everybody started raising hell.

"What are you doing, Rhinehart?" Mott inquired.

"I am making a new menu. I will have to revise all of the meals again. Our noon meal will have to be limited to that cup of dishwater the Germans give us. I have just figured the cigarettes, and it comes to twenty-one to each man for a week. There are seven men in our combine, so we get seven-twelfths of these three parcels. The remaining five-twelfths goes to Combine D. Jones asked me to divide them up."

"You may divide them," Tom said, "but don't worry about Jones in Combine D. He will check you to the last ounce of coffee. They tell me he uses a hack saw when he divides the chocolate ration."

"What is that?" Gene suddenly shouted.

A tremendous roar overhead shook the barracks.

"Fighter planes," someone shouted.

We hit the floor by the dozens. Swanson dived under Pirtle's bed, and Pirtle had to crawl under the table. Plane after plane roared over at rooftop level. Several goons in the towers fired rifle shots, but no fire was returned.

In a few minutes they were gone, and we heard everybody cheering outside. We let Swanson get into the hall, and we followed him outside.

Outside the cheers were loud and long.

"What happened, McGee?" John asked him.

"American P-47's just buzzed the camp. They weren't two hundred feet off the ground. I guess they wanted to cheer us up. Look, here comes Pop-Eye with Dumbkoff."

Pop-Eye was shouting amid cheers, bellows, backslapping, and laughter.

"Idiots!" he shouted. "Inside! 'Raus! Inside, American idiots!"

For the next several days we talked about our unexpected visitors. Pop-Eye had been so mad he locked us in for two hours. Mott got his bang of the week, not from the planes, but from Pop-Eye's anger.

By the end of the week the P-47 buzz job had been talked to

shreds. The suspense was steadily mounting all over the compound. A new phase in Kriegie life was beginning, and it was to be called the "Sweat-Out." We were sweating it out for the rest of the way. Good news would eventually come, and when it did, I knew we would become all the more tense, anxious, and nervous. The impatience grew steadily—and then the greatest news of the war came to us.

On March 29, Captain McGee asked us to remain in the barracks after lunch for a special bulletin to be delivered by Intelligence. None of us knew what it was about, but the tension mounted to a high pitch before Captain Ferrell arrived with the bulletin. Ferrell came into the barracks excited and out of breath. He stationed his guards on both barracks' doors, and read from the center of the barracks.

"It's good news, boys," he said as he opened the sheets of paper.

"The west front," Ferrell began as he got his breath.

"On March 20, General George S. Patton, Jr., and the Third army smashed through the defenses of Worms, Germany, and captured the city. His forces drove thirty miles up the west bank of the Rhine River and captured 30,000 German troops. He threw the entire strength of his army into action and brought up massive reinforcements. With pontoon bridges and heavy fire power, he crossed the Rhine at Mainz; and after a tremendous battle, he captured the city on March 27th. At that point his army was 150 miles from the gates of Mooseberg.

"Yesterday he continued his drive, with the German army in full retreat. Battles were going on all over the sector as Patton's armored divisions smashed forward. Late yesterday evening, Frankfort-on-the-Main surrendered to General Patton. From our map, Intelligence has figured the American Third Army to be 150 miles from Mooseberg. Colonel King warns that the goons are getting jumpy, and we should do nothing to antagonize them. They are alarmed at the advances made by General Patton, and we don't want any men to get hurt. It has been reported that Patton is giving his German prisoners a rough time. That's it."

Some of us moved outside so the guards would not become suspicious of some kind of inside meeting. To us, the report was the most magnificent news we had heard since the beginning of the war.

The tension mounted to a high pitch as several days passed by with minor gains along the Allied front lines. The sweat-out grew in intensity. Men chewed their fingernails, walked the parade ground at swifter paces, and made twice as many trips to the latrines. The more nervous men became, the more difficult they were to live with. Each man measured his words for fear of starting an argument, but sometimes violent arguments broke out.

On the night of April 6, Captain McGee came into the hall excited and breathless. We knew it was good news.

"The news is as follows," he told us. "General Patton's Third Army and General Patch's Seventh Army met after Patton captured Frankfort-on-the-Main. They are now moving side by side across our sector of the western front. On March 30th General Patch captured Heidelberg, located 110 miles from Mooseberg.

"On April 5th, General Patton and his Third Army continued their advance in our direction, capturing Osterburken. Osterburken is located approximately 96 miles from Mooseberg.

"The Russians captured Hanover on April 2nd, and Brunswick surrendered to the Russian Army on April 3rd."

Most of our news came from radios that had to be assembled by Intelligence. Sometimes the goons were too near at hand to chance assembling them. Other times interference made it impossible to pick up the news.

On April 13, we got a morning bulletin that shocked the entire camp as it had shocked the Allied world. Captain McGee read the report.

"Yesterday morning, President Franklin D. Roosevelt died at Warm Springs, Georgia. He died from a cerebral hemorrhage. Vice-President Harry S. Truman became President of the United States a few hours later. All of our nation is in mourning today."

Many of us could not believe this news at first.

"At ease," McGee shouted. "I have a message from Colonel King.

"'Gentlemen. It is with great sorrow and a sad heart that I tell you our beloved President is dead. By special permission of the commandant, all American compounds and the British and French compounds will hold a special memorial service at two o'clock today. Silver taps will be blown at the close of the ceremony. Pop-Eye will furnish one of the bugles to a member of our old Sagan band.'"

Promptly at two p.m. all American compounds and the British and French compounds were formed in the center of each parade ground. Colonel King called the Center Compound to attention. As he did so, we could hear other compound commanders calling attention.

Colonel King stepped forward and said: "A great man, and the leader of our beloved nation, died yesterday. We have gathered here today to pay our last respects. I think no greater tribute could be paid him than for us to say, as prisoners of war, that he was the greatest soldier of them all, and he gave his life for his country."

Chaplain Daniel prayed, and many men cried. I remember his closing words. "Greater love hath no man than this, that a man lay down his life for his friends. In my Father's house are many mansions; if it were not so, I would have told you. I go to prepare a place for you. Our President has found his place, and he is at peace forever. Amen."

Standing next to Chaplain Daniel, one of the members of the Sagan band raised the bugle to his lips, and sounded the minute-long "Silver Taps." The music rose into the air, and swept over the compounds of Mooseberg. German guards stood at attention. Pop-Eye removed his cap.

On April 21 the good news was almost beyond belief.

Captain McGee received the report from Captain Ferrell during lunch, and asked everybody to stop eating and listen.

"At ease. I have news more important than your lunch. On April 9, Rothenberg surrendered to General Patch. Rothenberg is 71 miles from Mooseberg.

"General George S. Patton and the Third Army captured Hall on April 11. Hall is 68 miles from Mooseberg."

We sat spellbound, for we felt that something big was coming.

"Following the capture of Hall, General Patton led an armored attack toward Ansbach. He captured an entire German division and encircled the city. On April 16, Ansbach surrendered to Patton. Ansbach is 45 miles from Mouseberg."

We nearly went wild with excitement. Old Kriegies like Mott were stunned by the report.

"At ease!" McGee shouted. "General Patch and the Seventh Army spearheaded a drive on Nuremberg. General Patton turned the Third Army to Nuremberg to join forces with General Patch. Yesterday afternoon, the Seventh and Third Armies captured the city of Nuremberg, and liberated 30,000 prisoners of war in the Nuremberg prisons. Nuremberg is 43 miles from us. The last news report said that General Patton and the Third Army are now advancing in the direction of Munich. That's it."

For the first time we disregarded all Kriegie regulations about the news. We cheered, hollered, shouted, jumped up and down, and got hysterical.

"Let's go outside and have a party," Gene said excitedly.

Seven of us rushed to the north end of the barracks, and got our favorite spot by the north wall.

"Well, boys," Clark said, "if nothing happened to them, Peter Glass, Dave Roberts, and Ted Carson are free men."

"That is the most wonderful news of our lives," Swanson said.

"I hope that all of them survived the death march. They didn't say anything about there being any casualties, so Patton must have taken the camp without a struggle," Mott remarked.

Most of us felt we would be out of Mooseberg within four days at the most. I know that our blood pressure rose as tension continued to mount among us. We were extremely excited, and several hundred men were pacing the north, east, and west fences looking for signs of Patton's Army.

During the next week, tension mounted to the limit as we talked, waited, and watched for something to happen. At every appell we could tell that Pop-Eye and all of the goon guards were getting more nervous. We gave Pop-Eye our fullest cooperation, not wishing to upset him.

On the twenty-seventh of April, Allied fighter planes started circling our camp. They came down as low as possible without getting into shooting range. We didn't cheer this time. Planes flew over all day, and Pop-Eye was about to explode with tension. We sat on the ground and looked, but nothing more. It was too near the end to have some German guard go crazy and start killing some of us.

At noon the commandant ordered a special appell. He came through the gate in full dress uniform with his staff and Rhoder. Pop-Eye and the guards took their positions beside the various blocks. The commandant had Pop-Eye read the following report to us.

"Attention! In the event this camp is attacked by the American Army, a loud siren will be blown from post headquarters. You will all move to your barracks at once. All barracks doors will be closed and locked. Any American prisoner giving us trouble of any kind will be shot. It has been decided that it will be impossible to evacuate this camp. It is further ordered that if American planes come over this camp again, you will also move immediately to the barracks. No siren will be blown, but any man disobeying this order will be shot. By order of the Commandant."

The commandant marched out of the compound, and we were dismissed.

That night we understood the commandant's alarm. Colonel King came into the barracks before lock-up time, and commanded "at ease."

"I have a special news bulletin. General George S. Patton and his Third Army captured Ellingen this afternoon. Ellingen is nine miles from Mooseberg."

No one shouted.

"General Patton dispatched a staff car under a white flag to Mooseberg post headquarters late this afternoon. He has asked that the commandant, the senior American prisoner of war, and his chief of staff of the Third Army meet at noon tomorrow at post headquarters at Mooseberg.

"The commandant informed me an hour ago that I was to attend the meeting with him. Men, I believe that General Patton will try to negotiate the surrender of the camp, in order to save as many of our lives as possible. What the commandant will decide is hard for any of us to say. If there is a battle for the camp, I will give you final instructions tomorrow. That's it."

XXIV

The Battle of Mooseberg

ON APRIL 28, IMMEDIATELY AFTER APPELL, COLONEL KING, SENIOR Allied Prisoner of War, left the compound with Pop-Eye. The next seven hours were the most anxious and tense hours of our prison life. We talked, walked, and paced the compound.

Finally, at four P.M., Colonel King was escorted through the gate by Rhoder. They talked a few minutes, and Rhoder departed. Everybody rushed toward him, but Colonel King declined to make a comment.

"I will issue a bulletin this evening." His face was stern; it didn't foretell good news.

At eight P.M. the report arrived. Captain McGee called us together and read it.

"My fellow prisoners. As you all know, I met at noon with the commandant, and Major General Smith, Chief of Staff of the Third Army. General Smith read a proposal from General George S. Patton, Jr., requesting the commandant to surrender the Mooseberg prison without combat. In return, General Patton would guarantee the German general, and his staff, and all German military personnel, that there would be no military trials for war crimes, and that all German personnel would be treated as prisoners according to the terms of the Geneva Convention. A colonel in the S.S. was also present, and at his insistence the general declined the terms and decided to make a last-ditch fight.

"Major General Smith then informed the German general that General George S. Patton's Third American Army and part of the American Seventh Army would attack Stalag 7-A at

8:00 in the morning, April 29, 1945. Patton warned that if any of us were harmed by the Germans, neither the commandant nor any of his guards would ever live to tell it.

"When appell is over tomorrow, you will wait for the siren, which will warn us of the beginning of the attack. Move to your barracks at once. When we have been locked in, listen for the first sound of bullets. Those of you who wish may rip out the floorboards and get under the barracks. Stay on the floor, or get flat on the ground. Most bullets will be flying about three feet from the ground. Stay flat, and do not move around trying to see the battle. That will be a good way to get killed. Steady your nerves, for this is the end of our imprisonment. Carry out these orders, men, and we will be free tomorrow before dinner. May God save all of us from harm. When I see you again, we will be free men. God bless you." (Signed) Colonel King.

So that was it. Our Mooseberg commandant, who was much under the influence of the S.S., was a fanatic and would not give up or surrender. Our Sagan commandant, a Luftwaffe man, lacked this fanaticism. But to Hitler's fanatics, like our present commandant, surrender was a disgrace.

Did the Germans have any hope of a successful defense of Mooseberg? It is possible that the commandant thought he could obtain concessions from General Patton by a decision to resist. He probably wanted to make Patton hesitate before the possibility of inflicting heavy casualties upon thousands of American prisoners of war. I am not sure that he reasoned that way, but I do know that the Germans feared most two things: the Russians and General Patton. They considered Patton ruthless, and this induced the will to resist, even if resistance was hopeless.

It was 7:25 A.M. on Saturday, April 29, when the German bugle sounded in the Center Compound for appell. Pop-Eye and all of the guards were waiting for us to fall in formation. They were all dressed in clean blue Luftwaffe uniforms. Both Germans and Americans were tense and stern. The appell was taken in a strained silence. The commandant received the report from Pop-Eye and departed from the compound.

Pop-Eye, Dumbkoff, Shisenkoff, Bear-Face, and the rest of the guards remained in the compound. We had eaten breakfast early so there was nothing to do but wait. Most of us used the latrines and walked around the fences. At 8:05 A.M. I was standing at the north fence looking toward the nearest hillside about two miles away. There wasn't a living thing in sight. At 8:10 the camp was quiet, but nothing could be seen.

It started at 8:14. Large clouds of dust rolled over the hill in clear view. Tanks appeared over the hill. I looked down the hillside as far as I could see, and tanks were lining up for miles and miles. I looked up and saw five squadrons of fighter planes coming our way. The siren sounded loud and clear over the entire camp.

I ran to the barracks with Gene, Pirtle, and Swanson. In a few minutes we were locked in, and all shutters were closed. Floorboards were quickly ripped from the floors, and three passages were made under the barracks. Men started getting under tables and beds; others lay on the floor. Seventy-eight of us climbed quietly under the barracks. The first under spread out to give others room. All of us lay on our stomachs.

The planes could be heard overhead. They were still high, and they were circling the camp. The earth began to shake and rumble from the movement of several hundred tanks. All of us were smoking except Gene. Several single shots were fired, probably from the sentry towers. Then the roar of fighters' engines rang in our eardrums. They were diving and racing toward the earth.

Machine-gun bullets began bursting in every direction. Several bullets whizzed overhead. Planes swept over the camp at ground level altitudes, with machine guns blasting. Swanson looked out from under the barracks.

"I can see them," he shouted. "They are attacking the sentry towers. Holy heaven, they aren't fifty feet off the ground. They are diving on the towers and firing single bursts of fire at the towers."

I chewed my cigarette, and bit into my lip. The tanks were getting close now, and the ground shook as they moved in our direction. The roar of tanks got louder, and German guards started shooting machine guns. The volume of fire seemed to increase from all directions. The roar of tanks, planes, and guns blasted against our eardrums. My ears were ringing like bells. I prayed and smoked harder than I had ever done either before.

Cold sweat broke out in the palms of my hands as the tanks

got close to us. Sweat popped out on my forehead, neck, and legs. I looked at Pirtle. Sweat was running down his face. His fists were clenched tight, as if he expected at any moment to feel the pain of a bullet. Clark was praying aloud with his face cupped in his hands. A bullet ripped through the barracks overhead, and we threw ourselves prone in the dirt. German guards were screaming, bullets were flying, tanks were shaking the ground like an earthquake, and diving planes were spitting fire.

All at once Dumbkoff bounced in, landed against Swanson, and crawled on top of us. We made room for him and he shouted: "Surrender, surrender." I was too scared to smile.

We heard the crashing and ripping of steel. The firing died out, and we couldn't hear the sound of fighter engines. Suddenly everything stopped except the movement of tanks close by. We heard three or four additional single shots, and then it got quiet. The tanks had stopped. We all looked at each other. My eyes and face were covered with sweat and dirt, but so was everyone's.

Someone screamed out: "It's over, we are free, the battle is over, everybody come out!"

We listened a second or two more, but we heard only American voices. Slowly we started moving from under the barracks. Dumbkoff handed his pistol to Clark.

When I got out from under the barracks, I looked about. Men were climbing out of windows. Kriegies were shoving their way through the doors, and many men were climbing to the roofs of the barracks. I looked toward the north gate and nearly fainted. There was an honest-to-God American tank parked ten feet from our barracks. American soldiers were climbing from its top. The barbed-wire fence behind the tank was smashed to pieces. Strands of wire were hanging from what remained of the fence. Hundreds of us rushed toward the hole in the fence.

Pirtle yelled: "American tanks, American soldiers. We are free! Thank God, we are free!" He broke down and cried. He wept bitterly as Clark put his arm around him and pushed him forward toward the tank. We rushed to the soldiers, and they rushed to meet us. We met at the north end of the barracks. We embraced each other and shouted and cheered. Pirtle put his arm around Swanson's shoulder and tried to stop crying.

The battle had lasted not quite twenty minutes. Only for three or four minutes had the fighting been fierce. The fighting had been concentrated on the pillboxes outside the barbed-wire entanglements and on the sentry towers. American planes had pinpointed and knocked out the outlying pillboxes and American tanks had knocked out the sentry towers and smashed the barbed-wire barriers. There were three separate series of barbed-wire fences; when all of these were smashed the battle was over. Inside the barbed wire, guards such as Dumbkoff had only pistols and they did not resist. Most of the guards surrendered when the tanks came through the fences. Our ground forces in trucks followed but there was no fighting for them to do.

We looked down the street and saw a tank rushing toward the American Compounds. It had crashed through the gates further west of us. It rolled to a stop ten feet from our barracks, and the top swung open. We expected to see more G.I.'s pour out. Instead a helmet appeared above the opening, and a large, broad-shouldered, stiff-backed soldier lifted himself from the tank. He stood on top of the tank and looked about. We could hardly believe our eyes. It was General George S. Patton, Jr., Commanding General of the Third Army.

It was General Patton himself with his stern face, sharp, piercing eyes, and rugged features. His helmet glistened with four brightly painted stars. He was wearing a G.I. uniform with G.I. battle jacket, and had a set of four silver stars on each collar and each shoulder. Around his waist were two pearl-handled pistols. He was wearing brown cowboy boots neatly shined. He looked up and down the street checking the location of his tanks as Kriegies swarmed around him. Then the rest of the members of the tank crew started climbing out.

Patton smiled down on us and shouted: "The war is over for you boys. You are sure going to eat tonight. I damn sure guarantee you that."

He climbed down from the tank and started shaking hands. He acted like Mr. G.I. Joe himself, and used both hands as he greeted the men. He moved toward the hole in the fence near us shaking hands, asking questions, frowning, and stopping momentarily from time to time. At one time he got within five feet of us. He spoke in a loud, commanding voice.

"You boys need some food and you are going to get it. How many sick men do you have in here?"

Some Kriegie told him three or four thousand.

"Don't you boys worry about a damn thing. I will see that you are cared for. I am going to inspect this camp, and I will get your sick men out of here. These German bastards are going to pay for this. The sons-of-bitches will suffer, and don't you ever forget it."

He passed by the barracks walking toward the parade ground. His cowboy boots glistened in the sun. He was immaculate, dignified, and as tough as his reputation.

He entered the barracks next to us, and as many as possible followed him inside. We stood at the open window and watched General Patton. He stopped at the third bed on the left side of the barracks next to us and bent over a sick Kriegie. His face was stern at first, and then he smiled.

"Son, I am George S. Patton, Jr., commanding General of the Third Army."

The boy looked up in disbelief. He looked the General over thoroughly and answered, "Really." He had been too sick to know what was going on.

"Don't you worry about a thing. The battle is over, and Mooseberg is liberated. You are safe, and I will take care of you. We will have you out of here before the day is over."

The General shook hands and proceeded along the line of tiers until he came to the next man in bed. Kriegies crowded and shoved against him as he spoke.

"I wanted to shake the hand of a brave man," Patton said.

The boy looked up and smiled, recognizing him from his pictures.

"God bless you, son. We have liberated this camp, and you are free. I have help on the way, and you will be out of here in no time."

When he started down the hall again, he was beyond our sight. No one tried to go inside as the barracks were jammed to capacity.

"General George S. Patton," I said in amazement. "Stars, pistols, boots, and all. What a man! What a soldier! What a heart he must have! I always heard he was hard as steel, but he is the kindest man I have ever known."

"Look!" Rhinehart shouted. He wiped the tears of joy from his eyes.

"Where?" Pirtle asked. All of us turned.

"Look at all those men on the roofs of the barracks—look at those characters climbing fences."

All of us started toward the parade grounds to see what was going on. American officers and soldiers of Patton's army were all over the camp. A group of six G.I.'s approached us.

"Hi, fellows," one of them said. "You boys have lost some weight. This is the worst hellhole we have seen in Germany. Most of the army is moving on, but we have been assigned to help get you out of here. Don't worry about a thing. We will have this camp reorganized tomorrow, and we will start feeding you."

They followed us out to the center of the parade grounds.

Everybody started shouting: "Look at the flag! Look at the flag!"

We stopped near the west fence which separated our compound from the British compound.

We turned with thousands of other men toward the north, in the direction of camp post headquarters. We saw the German flag slowly descending the flagpole. Silence swept over the camp. I turned my eyes to the British compound and saw men look in disbelief. English soldiers stood spellbound, breathless, absolutely quiet. They looked and watched as the German flag disappeared. I looked over my shoulder to the south and saw French soldiers remove their caps. I saw others looking, watching, and waiting. I looked back to the front.

It seemed like ages before it happened. At first men from the rooftops started cheering and shouting. Then thousands of voices rang out in the air.

Slowly, steadily, and completely unfurled, Old Glory came in sight and slowly ascended the flagpole. Tears from my eyes blurred the picture. The Stars and Stripes waved and rippled as they moved slowly toward heaven. Soldiers from General Patton's Army snapped to attention and saluted. Those of us who were able joined in with them.

When the flag climbed to its full height, salutes were concluded, but eyes remained on the Stars and Stripes. Men about me were kneeling. I knelt too. While kneeling, I prayed, and thanked God. I wiped my eyes, but the tears would not stop. Gene Coletti knelt beside other happy men and wept bitterly. Pirtle covered his face again. Clark stared in disbelief, Swanson blew his nose, Rhinehart rubbed his eyes and coughed. Mott held his head high. His eyes were filled with tears, but he smiled.

I turned my head to the west again, and looked through the fence. Men of the British Compound were weeping. Some were standing and looking in utter disbelief. Some cried aloud. Others knelt and prayed. I saw several men close to the fence sitting on the ground. They picked up hands full of dirt and watched it drop to the ground. They looked at the flag, they looked at us, they wept, and picked up more dirt. These men were stunned. Some of them had been prisoners of war since Dunkirk. They had suffered unknown agonies, and now they were supposed to believe and accept the fact that they were free men again.

The ceremony was spontaneous and individual. For some it lasted five minutes, but for others it lasted an hour. Officers and enlisted men of the Third Army who were with us stood in silence. Some cried, while others watched, stunned and amazed.

There wasn't a man among us who was embarrassed or ashamed. We walked arm in arm with the G.I.'s, talking and laughing. The war was over and we knew it. Freedom was wonderful; we could hardly realize that we were free.

Some of the guards in the sentry towers had been killed. The remainder of the guards in our compound had surrendered with Pop-Eye. They were being assembled at the north gate outside the compound. We moved to the gate and watched as G.I.'s marched goons to the formation. Pop-Eye and Dumbkoff were in the formation, but we never saw Shisenkoff or Bear-Face again. Pop-Eye looked sad, but Dumbkoff seemed to enjoy the whole affair. Many Kriegies got their revenge by shouting everything in the book at the defeated goon guards. The most popular shout was, "Swine, Luft gangsters, idiots. Idiots, Luft gangsters, swine." Pop-Eye stood in the formation with eyes straight to the front. Dumbkoff was scared, but you could tell he was glad it was over.

Finally a G.I. sergeant called them to attention and marched the formation off in the direction of post headquarters. As they started off, Mott rushed forward and shouted to Pop-Eye.

"Good-by, Pop-Eye. Tell the Fuehrer and Rhoder that we won't be there to help them rebuild Berlin. General Patton told us der Fuehrer is going to pick up bricks and stones every day. Dumbkoff is going to count for der Fuehrer, and every time he drops a brick Dumbkoff will kick him in the butt."

Mott had everyone laughing except the goons. They didn't think it was so funny. As we watched them march down the street, I wondered if Pop-Eye wasn't really glad that it was all over.

American Kriegies from the north compound across the street had joined some of us, and we exchanged stories. As far as we could learn, only one American prisoner was killed in the liberation. A man in their compound had rushed to the gate and had been shot trying to get over the fence. None of them knew who shot him.

Soldiers of the Third and Seventh Armies were pouring into the camp. The streets were jammed with trucks, jeeps, and soldiers. All of them were trying to find the American Compounds. Dozens of officers and soldiers were taking pictures of us.

George Swanson saw them first. He shouted: "Women! Look, they are alive. Real American women, breathing and moving toward us."

The crowd started cheering, and soon I couldn't hear a word anyone was saying. The first three truckloads of them were nurses and Red Cross workers. They were all in uniform, they were smiling, and we thought they were the most beautiful women we had ever seen. We climbed fences, stood on trucks, cheered and whistled as they drew near. All of us were laughing, and some of us were hysterical. Two canteen trucks pulled up in front of our door, and more girls got out. They started passing out gum and cigarettes. It was announced that coffee would be served in a few minutes.

A sound truck with a loudspeaker started playing records. The first American song we heard was *Don't Fence Me In*. In a few minutes all of us were smoking, talking, singing, chasing women, or standing in line for coffee. Each girl had two or three

G.I.'s helping her, and I wondered if they would have braved the crowd of wild Kriegies without an escort.

This was the greatest celebration we would ever know, and we were "living it up." Colonel King appeared at the gate, and we gave him a thunderous welcome. We sang *For He's a Jolly Good Fellow*. We knew he had been with General Patton, and a path was cleared for him to the street.

He made his way to the sound truck and said: "At ease, men." Everybody began to quiet down.

"First of all, I would like to thank you soldiers of the Third and Seventh Armies who liberated us this morning." Loud cheers drowned him out. He continued: "It goes without saying that this day will live in our memories forever. I think you are aware of our feelings, and we are eternally grateful to you. Now, men, you can visit and stay out here for another hour. When the whistles blow, I want all of you to return to your barracks. The sick men must be removed. General Patton's men are trying to get food, clothing, and medical supplies into this camp. Plans for processing you are being drafted, and we want to get started this afternoon. Your block commanders will give you special orders after lunch. We will be here several more days, until the processing can be completed and General Patton can make the arrangements to move us out. I want you to give your fullest cooperation to those military personnel in charge of getting us home."

All during the day supplies were brought into the camp. Clothes, blankets, food, doctors, nurses, and specialists were everywhere. By mid-afternoon every bedridden prisoner had been removed from the Center Compound. Some were rushed to nearby hospitals, and some were taken to special landing strips and flown directly to France. Portable kitchens were moved into every compound, and G.I. cooks started to work.

At three o'clock, Captain McGee read a special directive.

"I have been appointed by General Patton to handle the evacuation of Mooseberg. It will be several days before we can start moving you. Americans will be removed first. British, French, and Russian officers are being flown here to handle the evacuation of their own men. We will start feeding you tonight. My staff surgeon has informed me that all of you must remain on a soft diet for several days. Your stomachs have shrunk, and they must be stretched by degrees. The doctors warn that overeating or eating solid foods could be very harmful to any of you. Please comply with all orders so we may move you out of here with the least amount of delay. We have two cards for you to fill out, and one letter for each of you to write home. One card is for the Red Cross, so your next of kin can be notified immediately. The other card will be flown to France to your next base so they can prepare for you. Your letter will be flown to the United States by military aircraft. Tomorrow we will start cleaning you up." (Signed) Major General Jones.

At 5:00 P.M., the bugle sounded for chow, and we headed for the line. Each of us got a field mess kit with knife, fork, and spoon. As we passed by the field kitchen, we were served creamed potatoes, soft scrambled eggs, and milk. The mess kit was filled, and it was more than we could eat. However, some of us went back for seconds on the milk.

During the meal it was announced that General Patton would address us before his departure. We gathered in the streets at the north end of the compound. General Patton spoke over the loud-speaker, standing on top of a G.I. truck. He had spent the entire day in camp shaking hands with every American prisoner who was unable to stand or walk. He looked tired, stern, and sad. His speech was short and to the point.

"These bastards started this war, and we finished it. It will be over in a few days, and you will be going home. You men have displayed real bravery and courage. You have survived a great ordeal. Some of you are gravely sick, but I intend to bring every available doctor into this camp. Now I must depart to the front lines, if there are any left. I am proud that I had a part in winning this war, but most of all I am proud of liberating you men. Goodbye, and God bless you."

He rode away in a staff car with stars, boots, and pistols shining.

We cheered until he was out of sight.

Epilogue

THE EMOTION WE FELT WHEN THE AMERICAN FLAG ROSE ABOVE Mooseberg was never surpassed. This was the climax, the highest point, of our prison experience, and what followed was in the nature of epilogue. Here, in a few pictures, is what happened between our liberation and our resumption of civilian life at home.

•   •   •

In the next few days hundreds of General Patton's trucks were stripped and dismantled for hauling troops. On May 7 the truck caravan started transporting American prisoners of war to airfields and specially built landing-strips at Augsburg, Germany. We packed our belongings, and prepared to depart. In my collection were seven rolls of toilet paper. In these seven rolls I had concealed thousands of words, a diary of our entire experience.

Combine C was evacuated in the second convoy of trucks.

As the convoy started through the front gate, Mott said: "Goodbye, Pop-Eye and Dumbkoff. Goodbye, Stalag 7-A. May we never see any of you again." He expressed our sentiments to the letter.

We arrived at a large Luftwaffe airbase outside the city. Hundreds of engineers had been working repairing and building runways for planes to evacuate us. When we got there, we witnessed a massive operation. Dozens of C-47 transports were landing and unloading supplies. Then they reloaded with twenty-five American prisoners of war aboard each plane, and took off for France. We watched, and sweated it out for hours.

Finally, just before dark, Captain McGee announced that we were to spend the night in the barracks on the base and that we would leave in the morning. He told us we were being flown to Le Havre, France, to Camp Lucky Strike. From this camp we were to go home by boat.

We went to bed early, as most of us were tired from standing and waiting for a plane all day.

On Monday morning, May 8, 1945, we again got our equipment and lined up on the long runways. Our group was to board the tenth plane.

All at once the sirens from the control tower started blasting away. All of us turned and looked at the sky. Then a voice spoke over the loud-speaker.

"Attention, all men! Attention, all men! The war has ended! The war in Europe is officially over. The German Armies have signed an unconditional surrender. General Eisenhower has accepted the unconditional surrender, and the war is officially over."

Loud cheers and roars could be heard all over the base. We cheered, laughed, and jumped up and down. After five long years of conflict the war had finally come to an end. It was not only over for us, but it was over for all soldiers in Europe. Now to finish off Japan, and the world would once again be at peace.

While we were talking and cheering, a single plane flying at about two thousand feet altitude approached the landing strips. Suddenly we turned to look, and recognized a German Stuka dive bomber. The plane's nose dipped as she dived toward us at full speed. The sirens started blowing, and we scattered faster than I had ever seen a group of men run. Hundreds of us dropped everything we were holding, and rushed in the direction of hangars, buildings, trees or anything that could be used for cover. Men hit the bare ground and tried to cover in the shelter of the grass in the fields. Swanson fell on top of Pirtle and me. The plane came roaring down on us, and opened fire as she started pulling out of the dive. Machine guns from the towers and patrol cars opened up in every direction. As the Stuka started pulling up, a dozen machine guns spurted fire all about the plane. She rose to about two hundred feet and exploded in flames. The remnants of the Stuka plunged toward the woods at the end of the runway where they fell flaming through the trees. Every one of us hugged the ground and looked to the sky for more planes. None were in sight, and none came over. We stayed on the ground for several minutes as the control towers radioed news of the attack. Several minutes later, two squadrons of P-47's appeared over the base at low altitudes. We started getting up, still numb from the shock. Fire trucks rushed to the wooded area to put out the fire.

In a few minutes the loud-speakers blared out: "All clear, all clear. It was a single plane. They must not have known the war is over. Those planes in the air are our fighters, and we will have protection up there as long as it is deemed necessary."

"I think it was just a stray Stuka," Rhinehart told us. "We probably won't see any more, especially with those fighters up there guarding us."

"Well, I certainly hope not," Gene said. "That would be the crowning blow if some of us got killed on the way home."

"Look!" yelled Clark. "There are some planes coming out of the north."

This time we started for the hangars with tremendous speed. We never tried to look back or identify the planes. Swanson's long legs passed every member of the combine. Before we got across the field to the hangars, the loud-speakers came on again.

"The planes approaching the field are American C-47's. They will land and unload.

"When each plane is ready for takeoff, we will call the number from the tower. You have the number of the plane you are to board. You will carry your belongings and move at once to the gangplank for immediate loading and departure. We ask your full cooperation, as we intend to move several thousand of you out of here today."

We started walking to plane number ten where she had parked. Trucks and crews were unloading her when we arrived. Captain McGee was in our group of twenty-five. He lined us up and checked our numbers. The pilot of the plane came forward and introduced himself to each of us. He shook hands, and told us it would be a two-hour flight.

• o •

At Le Havre the trucks pulled out on schedule, and we waved goodbye to the French civilians who had cheered our arrival.

We drove along a coastal highway that overlooked the English Channel. The weather was warm and the sun gave us a feeling of comfort. Several miles out of Le Havre we saw Camp Lucky Strike, an American army base originally designed like an average American base in the United States. It covered several miles of land, and was a city within itself.

We finally came to a long row of buildings at the end of the base.

Swanson saw the sign first.

"Delousing," he grumbled. "Not again. Hell, you would think we had bugs or something."

We lined up outside to listen to the first doctor. We didn't know it at the time, but before we left Lucky Strike we were going to see more doctors in two weeks than most of us would see the rest of our lives.

After we had bathed and powdered, we moved down the hall to the clothing department. You didn't have to worry about size —you were issued a whole wardrobe in three minutes and moved down the line to the dressing room at the other end of the building.

On the fourth day we finished our medical examinations at Camp Lucky Strike, but the processing continued. We were gaining better than a pound a day, and most all of us were feeling wonderful. The average Kriegie had lost from thirty-five to forty-five pounds.

On the fifth day we went through the Records, Intelligence, and Information processing centers. We filled out dozens of forms and answered hundreds of questions for Intelligence. Gene and I reported about our crew-members. We listed Hollis, Bulla, Wheatly, Poltra and Falla as dead. We told them that Mather, Alston, and Tim had left Dalag Luft ahead of us, and that we had seen Alston at Weissen.

Several days later we checked with Intelligence Information, and they had records of only Tim, Gene, and me in their files. They assumed that Norman Alston and John Mather had died sometime during their imprisonment. They informed us that we would be notified at home when the information had been compiled. We were deeply saddened over their probable death in prison. We were naturally interested in finding out how they died, but neither of us could get very interested in reading a report on the account of their deaths when we got home. I wanted only to remember them as I remembered them when they were happy and alive.

Swanson and Clark also learned from Intelligence Information that Peter Glass and Ted Carson had been liberated at Nuremberg, and they were at another hospital camp in France. Dave Roberts had died in Nuremberg from pneumonia.

For the next few days we spent our time at the officers' club, reading, playing pool, poker, and bridge, listening to good music, and drinking malts between meals. All of us were getting fat as pigs. It was good to be in an American military camp and know that we were going home.

One day in the club Gene told me: "Whatever happened to Tim at Dulag Luft is beyond me. I feel sure that he did his best, and that he did nothing wrong."

"So do I, Gene, and that is why I said it is best to forget the whole thing. The war is over for all of us."

"Ken, from the reports we have gotten, seven men on our crew and three men out of our original combine at Sagan died in this mess. Do you realize that is about half of the men we were most closely associated with since we came to Europe?"

"Yes, Gene, we have been very fortunate to survive this whole ordeal. I almost feel as if we are living on borrowed time."

On Wednesday, May 24, 1945, we gathered our belongings for departure to Le Havre. From Le Havre we were to board a hospital ship for the United States. We were all excited because this was the final leg of our long journey home. While we were getting packed, Colonel King came into the barracks and told us goodbye.

"I wanted to tell you about General Vandermann and Colonel Spivey. They escaped across the border into Switzerland as we were told at Mooseberg. They went to Eisenhower's headquarters with the captured German generals and made a full report of our imprisonment, the march of death, and our new locations at Mooseberg and Nuremberg. They gave valuable advice about our needs and conditions, and they helped in the preparation of rehabilitation centers such as Camp Lucky Strike. Colonel Spivey is still in Europe. General Vandermann flew to Washington several weeks ago to make a full report about us. He also made special recommendations about our leave, pay, and processing, and discharge when we arrive in the States. I think you will all agree that this camp has been extremely efficient and well-organized. General Vandermann is the man responsible for most of it, and I knew you would want to know. Good luck to each of you, and may you find an abundant life in America."

* * *

The ship was a medium-size passenger ship. It had three decks and three flights of staterooms below. Three hundred of us walked up the gangplank to the main deck and looked about. It was armed with anti-aircraft batteries and two medium-size revolving turrets. Sailors, doctors, and officers of the ship's crew greeted us and talked about the war. We were the first prisoners of war they had seen, and we answered hundreds of questions.

Finally a loud-speaker blasted, and we heard someone shout: "Now hear this. Now hear this. I am Captain Robertson. Welcome aboard the *Liberator*. We have converted to a hospital ship to carry you P.O.W.'s home. We renamed our ship for this mission. You will be assigned to staterooms, and we will depart from Le Havre in one hour. We have doctors and attendants to look after you. We have a wonderful supply of food, and we intend to see that you eat like kings. We have movies every afternoon and night. We have reading rooms, game rooms, and many other facilities. You will be required to take exercise every morning. The doctors will check you from time to time. Most of your time will be free to do as you please. We shall do everything in our power to show you a good time and a pleasant journey. The trip will take eleven days. Your petty officers in charge will now assign quarters."

We unpacked our belongings, looked the floor over, and returned to the main deck. We watched them hoist anchor, and we waved goodbye to the scattered French civilians. It was a glorious feeling to watch the *Liberator* slip out into the English Channel, and head for the deep blue Atlantic. We bent over the starboard side and watched her cut through the masses of water below. Le Havre got smaller and smaller. We moved up the English Channel along the coast of France.

The mess hall was large and very nicely furnished. The tables seated four, six, and eight men. Combine C sat at a table for eight. There were no chow lines for us to worry about. Our food was brought to the tables on trays, and we had merely to holler for a refill. Most of us had gained fifteen pounds since our liberation, and our stomachs were just about normal.

The meals aboard the *Liberator* couldn't be beat. We showered, shaved, and hurried to the mess hall. Having been starved for months in German prison camps, we were hearty eaters. During the entire voyage, I never saw food left on a plate at my table. Everybody ate a good meal, but all of us were extremely conscientious about not wasting food. If anyone had two bites more than he desired, it was eaten just the same.

Starting June 1, we looked for some sign of land several times a day. The Captain continued to tell us that it would be June 4 before we reached Boston, but we continued to look for land.

On the morning of June 4, we were busy packing, getting our clearance slips from the doctors, and saying goodbye to fellow Kriegies and the ship's crew.

I had just walked out on deck, a few minutes after ten, when I heard someone shout: "I see land!"

The rest of us rushed to the port side of the ship. At the end of the horizon we saw small brown specks. As the ship moved steadily forward, cutting its way through the green masses of water below, the brown specks began to take shape. The specks changed into a solid form of brown. We stood spellbound as the shoreline appeared. Buildings took their shape, and we recognized a modern city before us.

"If Dumbkoff could see this, he would go crazy," Mott said.

"Can you picture Dumbkoff, Shisenkoff, and Pop-Eye standing here with us watching the miracle of the twentieth century? We are looking at the United States of America."

It was a mad scramble to the staterooms. Most of us struggled with our belongings to the main deck. All sailors not on duty assisted those with the heaviest loads. Many of us were almost hysterical as we came out on deck again. We had to stand three deep at the port side of the railings.

Part of the docks were roped off, and the civilians filled every inch of space beyond the ropes. Several thousand people were on hand waving flags and cheering. Many of them carried banners and placards which read "Welcome Home." Two bands were playing the Army Air Force Song.

When we reached the gangplank, my throat was dry and thick.

All of us smiled and waved at the civilians who had come to greet us. When half of us were on the main deck, the Air Force band dipped its colors and started playing *The Star-Spangled Banner.*

Kriegies still aboard ship came to the railings; those on the gangplank stopped, and all of us faced the colors, came to attention, and saluted. All other men in uniform joined us, and the civilians lowered their hats over their hearts and faced Old Glory. I had heard our National Anthem played many times, but never like this. The music rang out in our ears as our eyes filled with tears.

When the band stopped playing, Byron Clark knelt on the ground and kissed the bare earth and concrete. Pirtle, Mott, and Rhinehart knelt behind him, facing the Atlantic Ocean. As I dropped to my knees, a hundred other men dropped with me.

Half of the civilians knelt, not knowing what to expect.

The amazed members of the Air Force band started playing *America*. The words rang out in my ears. "My country, 'tis of thee, sweet land of liberty, of thee I sing."

o   o   o

Following arrival in the United States, we P.O.W.'s were regrouped by home addresses. I was assigned to a group that was sent to Fort Sam Houston, near San Antonio. Here we were immediately given ninety-day leaves to go home.

While my leave papers were being issued, one of my best friends, Marshall L. Felker, Jr., younger brother of the pilot to whom I have dedicated my Kriegie story, arrived from Austin, which is about 75 miles from San Antonio. Two hours after my leave was issued, Felker, who was an Aviation Cadet at the University of Texas, and I were on our way to Austin where the University is situated. I had decided to spend three days in Austin en route to my home in East Texas.

We couldn't get a room right away at the Stephen F. Austin Hotel at Austin, and left Felker's car unlocked in front of the hotel while we waited in the Coffee Shop of the hotel. It was mid-afternoon when we were finally assigned a room and proceeded to the car to get my B-4 bag and uniforms. Everything in the car was gone! My consternation was great because my diary written on seven rolls of toilet paper had been secure all these days in the B-4, and now at the very end of the trail, it had been filched. It was the only thing I really cared about in the stolen effects.

My fraternity chapter, Phi Kappa Sigma, at Austin, the city police, and the fire department hunted for two days through the garbage disposal system of the city. It was thought that whoever had stolen my clothes would certainly throw away the rest of the contents of the B-4 bag; who would keep seven rolls of toilet paper? But my diary was never found.

Fortunately, from time to time I had flashes of total recall and I commenced the writing of this book. After a period of seven years a large part of the book was finished, but I could not completely feel the absolute naked truth of the life we lived as Kriegies.

In September of 1946 I awoke in the middle of the night in a cold sweat. I felt a total compulsion to complete the book and I knew at that moment that in order to make it absolutely true that I would have to relive my entire experience as a Kriegie. The next day conferring with my wife and business associates and friends, I decided on a starting date and location where I could be completely isolated. I went to a house on a lake. I worked out a detailed schedule of operations for my wife, household and my business from Monday morning through Saturday evening and made agreements with all parties concerned that I would return each Saturday night and Sunday.

On the scheduled date I moved clothes, food, typewriter, pencils and reams of paper to this lakeside house three miles from Avinger, Texas.

This experience was an hour-by-hour process of total recall in which time swept backward. It was necessary for me to start in a cold sweat for I knew I must relive the entire experience. It was thus that I began rewriting this book. There were times when I was so entranced with the work that I typed and worked all through the night and into the day, cooking, eating, sleeping and relaxing close to nature.

When you read this, if you laugh you can be assured as you read that I laughed as I wrote, if you feel sorrow I felt sorrow, if you feel joy and happiness, I felt joy and happiness and if you cry, know that I cried more than you in order to write it.

All P.O.W.'s suffered. There were moments of joy and sorrow, of love and hate. There were days of great hope and other days sunk in despair. Courageous endurance and a determined will-to-live were necessary for survival. Ingenuity, sacrifice, and tolerance had to be developed by every prisoner. Minds grew alert and sharp. Strong characters were developed under trying conditions, and, most important of all, God became known to every man.